Philip Alexander Knobel

Translesion DNA Polymerases

AF060833

Philip Alexander Knobel

Translesion DNA Polymerases

Targeting Translesion DNA Synthesis in the Context of Cancer Therapy

Südwestdeutscher Verlag für Hochschulschriften

Impressum/Imprint (nur für Deutschland/only for Germany)
Bibliografische Information der Deutschen Nationalbibliothek: Die Deutsche Nationalbibliothek verzeichnet diese Publikation in der Deutschen Nationalbibliografie; detaillierte bibliografische Daten sind im Internet über http://dnb.d-nb.de abrufbar.
Alle in diesem Buch genannten Marken und Produktnamen unterliegen warenzeichen-, marken- oder patentrechtlichem Schutz bzw. sind Warenzeichen oder eingetragene Warenzeichen der jeweiligen Inhaber. Die Wiedergabe von Marken, Produktnamen, Gebrauchsnamen, Handelsnamen, Warenbezeichnungen u.s.w. in diesem Werk berechtigt auch ohne besondere Kennzeichnung nicht zu der Annahme, dass solche Namen im Sinne der Warenzeichen- und Markenschutzgesetzgebung als frei zu betrachten wären und daher von jedermann benutzt werden dürften.

Verlag: Südwestdeutscher Verlag für Hochschulschriften GmbH & Co. KG
Heinrich-Böcking-Str. 6-8, 66121 Saarbrücken, Deutschland
Telefon +49 681 37 20 271-1, Telefax +49 681 37 20 271-0
Email: info@svh-verlag.de

Approved by: Zurich, University of Zurich, Diss., 2011

Herstellung in Deutschland:
Schaltungsdienst Lange o.H.G., Berlin
Books on Demand GmbH, Norderstedt
Reha GmbH, Saarbrücken
Amazon Distribution GmbH, Leipzig
ISBN: 978-3-8381-2052-2

Imprint (only for USA, GB)
Bibliographic information published by the Deutsche Nationalbibliothek: The Deutsche Nationalbibliothek lists this publication in the Deutsche Nationalbibliografie; detailed bibliographic data are available in the Internet at http://dnb.d-nb.de.
Any brand names and product names mentioned in this book are subject to trademark, brand or patent protection and are trademarks or registered trademarks of their respective holders. The use of brand names, product names, common names, trade names, product descriptions etc. even without a particular marking in this works is in no way to be construed to mean that such names may be regarded as unrestricted in respect of trademark and brand protection legislation and could thus be used by anyone.

Publisher: Südwestdeutscher Verlag für Hochschulschriften GmbH & Co. KG
Heinrich-Böcking-Str. 6-8, 66121 Saarbrücken, Germany
Phone +49 681 37 20 271-1, Fax +49 681 37 20 271-0
Email: info@svh-verlag.de

Printed in the U.S.A.
Printed in the U.K. by (see last page)
ISBN: 978-3-8381-2052-2

Copyright © 2011 by the author and Südwestdeutscher Verlag für Hochschulschriften GmbH & Co. KG and licensors
All rights reserved. Saarbrücken 2011

Table of contents

1. Zusammenfassung 3
2. Summary 4
3. Introduction 5
3.1. DNA damage overview 5
3.2. DNA damage response (DDR) 6
4. Mammalian TLS Polymerases: state-of the art 10
4.1. History and Discovery 10
4.2. Fidelity and activation of TLS 11
4.3. TLS Polymerase families 14
4.3.1. Family A: TLS Polymerases theta (θ) and nu (ν) 14
4.3.1.1. TLS Pol θ 14
4.3.1.2. TLS Pol ν 15
4.3.2. Family B: TLS Polymerase zeta (ζ) 15
4.3.2.1. TLS Pol ζ 15
4.3.3. Family X: DNA Polymerase beta (β) and TLS Polymerases lambda (λ) and mu (μ) 19
4.3.3.1. DNA Pol β 19
4.3.3.2. TLS Pol λ 20
4.3.3.3. TLS Pol μ 21
4.3.4. Family Y: Rev1, TLS Polymerases eta (η), kappa (κ) and iota (ι) 21
4.3.4.1. Rev1 22
4.3.4.2. TLS Pol η 23
4.3.4.3. TLS Pol κ 23
4.3.4.4. TLS Pol ι 24
4.4. Translesion synthesis 24
4.4.1. One- and two-Polymerase mechanism 24
4.4.2. Apurinic/ apyrimidinic (AP) sites 25
4.4.3. 7, 8-dihydro-8-oxoguanine (8-oxo-G) 26
4.4.4. Thymine glycol (Tg) 26
4.4.5. (6-4) photoproduct 27
4.4.6. Cyclobutane pyrimidine dimer (CPD) 27
4.4.7. Benzo[α]pyrene-guanine (BP-G) 28
4.4.8. Intrastrand-crosslinks 28
4.4.9. Interstrand-crosslink (ICL) 28

4.5. Relevance of TLS Polymerases in cancer therapy 29
5. References 31
6. Aim of the thesis 48
7. Inhibition of *REV3* expression induces persistent DNA damage and growth arrest in cancer cells 49
7.1. Abstract 50
7.2. Introduction 51
7.3. Material and Methods 53
7.4. Results 55
7.5. Discussion 61
7.6. References 65
7.7. Figure Legends 71
7.8. Supporting Information 76
7.8.1. Supplementary Figures 76
7.8.2 Supplementary Materials and Methods 80
8. Acknowledgements 85

1. Zusammenfassung

Heutzutage basiert die Therapie des Lungenkarzinoms und des Pleuramesothelioms meist auf Cisplatin-haltiger Kombinationstherapie. Bei der Mehrzahl der Patienten führt dies zu einer vorübergehenden Hemmung des Tumorwachstums, mit der Zeit entwickelt sich aber eine Resistenz des Tumors gegenüber Cisplatin. Studien in Modelsystemen zeigen, dass die Translesionssynthese (TLS) eine entscheidende Rolle in der Entwicklung der Cispatinresistenz spielt. Das ursprüngliche Ziel dieser PhD Arbeit war die Rolle von REV3, der katalytische Untereinheit der TLS Polymerase zeta (Pol ζ), in der Resistenzentwicklung genauer zu charakterisieren.

Die Hypothese, inwiefern die Inaktivierung der *REV3* Exprimierung die Cisplatin-induzierte Mutationsrate beeinflusst, wurde überprüft. Dabei wurde entdeckt, dass die Inhibierung der *REV3* Exprimierung auch ohne Zugabe von Cisplatin zu einem reduzierten Wachstum von Krebszellen führt, wohingegen Zellen kultiviert von nicht-malignem Gewebe deutlich weniger beeinträchtigt sind. Zu diesem Zeitpunkt wurde entschieden, diese Beobachtung zu verifizieren und die Ursachen zu entschlüsseln.

Es wurde in mehreren verschiedenen Krebszelllinien bestätigt, dass die Inaktivierung der *REV3* Exprimierung in einer Reduktion des Zellwachstums resultiert. In der Folge wurde gezeigt, dass die Inhibierung der *REV3* Exprimierung zu einer Akkumulierung bleibender DNA Schäden in Krebszellen führt und zu einer ATM-abhängigen Aktivierung des sogenannten DNA Schaden Reparatur Signalweges führt. Die Inaktivierung der *REV3* Exprimierung in Krebszellen mit p53 Wildtyp lässt die Zellen im G_1 Zellzyklus arretieren und induziert Seneszenz. Dies wurde durch die Anhäufung des Zellzyklus-Hemmers p21 und einer erhöhten Seneszenz-assoziierten (SA)-β-Galaktosidase Aktivität verifiziert. In Krebszellen mit fehlender oder reduzierter p53 Funktion resultiert die Inhibierung der REV3 Exprimierung in einer Hemmung des Zellwachstums und in einem G_2/M Zellzyklusarrest. Ein kleiner Teil der Zellen wird nicht arretiert, was zu einer numerischen Chromosomenaberration, der sogenannten Aneuploidie führt.

Somit konnte mit diesen Resultaten zum ersten Mal belegen werden, dass die Hemmung einer TLS Pol *per se* zu einer spezifischen Beeinträchtigung von Krebszellen führt. Demzufolge wird postuliert, dass die Hemmung von REV3 ein möglicher Ansatz für eine Krebs-spezifische Therapie darstellen könnte.

2. Summary

Today, the therapy of lung cancer and malignant pleural mesothelioma is based on cisplatin as part of a multimodality approach. Despite an initial response to chemotherapy in a proportion of patients, all will ultimately develop chemotherapy resistant disease. Studies done in model systems indicated that functional translesion synthesis (TLS) is a major contributor to the development of cisplatin resistance. The initial aim of the PhD thesis was to elucidate the involvement of REV3, the catalytic subunit of TLS Polymerase zeta (Pol ζ) in the development of chemotherapy resistance.

Surprisingly, while investigating how inhibition of *REV3* expression affects cisplatin-induced mutagenesis, it was found that inhibition of *REV3* expression *per se* suppresses colony formation of cancer cells whereas cells cultured from non–malignant tissue were less affected. At this time, it was decided to further investigate these findings.

Further, it was confirmed that inhibition of *REV3* expression in various cancer cell lines leads to a growth inhibition. In addition, it was shown that inhibition of *REV3* expression leads to an accumulation of persistent DNA damage, subsequently leading to the activation of the ATM-dependent DNA damage response (DDR) cascade. The inhibition of *REV3* expression in *p53*-proficient cancer cell lines results in a G_1-arrest and induction of senescence as indicated by the accumulation of p21 and an increase in senescence-associated (SA)-β-Galactosidase activity. In contrast, the inhibition of *REV3* expression in p53-deficient or p53-knockout cancer cells results in a growth inhibition and a G_2/M-arrest whereas a small fraction of the p53-deficient or p53-knockout cancer cells can overcome the G_2/M-arrest, which results in mitotic slippage and aneuploidy.

It is shown for the first time that inhibition of a TLS Pol confers synthetic sickness/lethality specifically in cancer cells. These findings reveal that inhibition *REV3* expression *per se* suppresses cancer cell growth whereas normal cells are less affected, thus identifying REV3 as a potential target for a cancer-specific therapy.

3. Introduction

3.1. DNA damage overview

Genomic information is stored as deoxyribonucleic acid (DNA) in every living organism and needs to be protected and maintained to guarantee genomic integrity. Each of the 10^{13} cells of the human body contains 30000-40000 genes encoded by 3×10^9 base pairs of the DNA [(1, 2) and reviewed in (3)]. The integrity of the DNA is constantly threatened either by spontaneous decay or by damage induced by endogenous and environmental sources, respectively. In addition, the accurate doubling of the DNA is an essential step carried out by the DNA replication machinery but errors during this process can also compromise the genomic integrity. For example, damaged DNA, which cannot be replicated by the high fidelity replicative DNA Pols, can lead to stalled replication forks and subsequent replication fork breakdown results in chromosomal instability.

The most frequent DNA alterations by endogenous sources result from spontaneous depurination/ depyriminidation processes and deamination of the base cytosine to uracil. Oxidative stress can lead to the generation of reactive oxygen species (ROS), which induce base modification such as 7, 8-dihydro-8-oxoguanine (8-oxo-G) and Thymine glycol (Tg) [reviewed in (4, 5)].

The most abundant environmental source of DNA damage is ultraviolet light (UV), which can induce nucleotide dimerization such as (6-4) photoproducts and cyclobutane pyrimidine dimers (CPDs) [reviewed in (6)].

The chemical agent cisplatin used for therapeutical treatment of lung cancer and malignant pleural mesothelioma forms DNA intra- and interstrand cross links (ICL), which can lead to a blockage of the DNA replication machinery (7). Benzo(α)pyrene is a major compound of tobacco smoke and forms, upon metabolic activation, a covalent Benzo(α)pyrene-guanine (BP-G) DNA adduct, which frequently mispairs during DNA replication with adenine therefore leading to guanine:thymine transversions (8).

To counteract the tens of thousands of lesions per genome per day, cells evolved a complex and interplaying system, the so-called DDR [reviewed in (9, 10)]. During DDR, DNA lesions are detected, leading to the activation of a signal cascade resulting either in the repair or the tolerance of the DNA damage, thereby regulating cell faith after genomic insult [reviewed in (4)] (Fig.1).

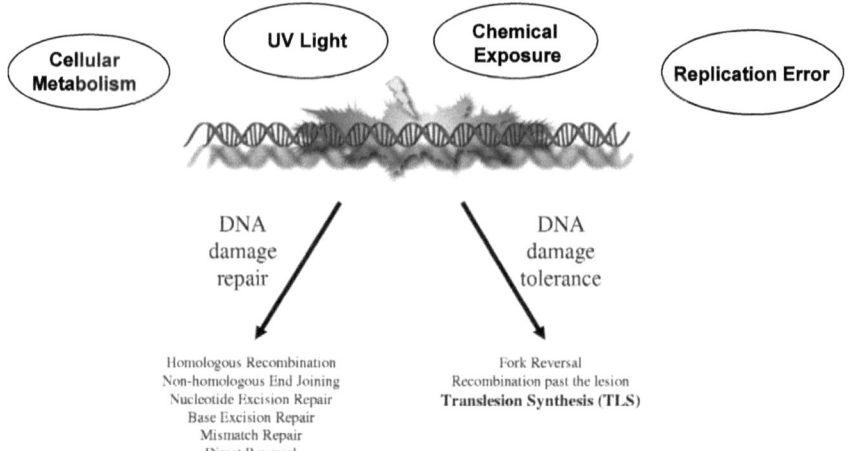

Fig.1 DNA damage induced by spontaneous decay or endogenous and environmental sources can either by repaired or tolerated [Adapted from reference (11)]. For details see text.

3.2. DNA damage response (DDR)

DNA damage induction results in the activation of a molecular cascade, e.g. the DDR, which consists of DNA damage sensors, transducers and effectors.

Induction of double-strand breaks (DSBs) triggers the ataxia telangiectasia mutated (ATM)-dependent DDR (Fig.2) whereas the formation of replication protein A coated single-stranded DNA (ssDNA) leads to the activation of the ataxia telangiectasia and RAD3 related (ATR)-dependent DDR (Fig.3).

Upon DSB induction, DNA damage sensors including the MRN (MRE11, NBS1 and RAD50) complex and the KU70/80 dimer activate the phosphoinositide-3-kinase-related protein kinases (PIKKs) ATM or DNA-dependent protein kinase catalytic subunit (DNA-PKcs), respectively [reviewed in (12)]. The activated ATM homodimer undergoes autophosphorylation leading to the dissociation of the homodimer (13) and the subsequent phosphorylation of the histone variant H2AX at serine 139 (γH2AX) adjacent to the DSB (14). Phosphorylated γH2AX recruits the mediator of DNA damage checkpoint 1 (MDC1) [reviewed in (15)], breast cancer 1 (BRCA1) [reviewed in (16)] and p53 binding protein 1 (53BP1) [reviewed in (17)] thereby accumulating additional MRN complexes resulting in a positive feedback loop, e.g. increased local ATM activity and γH2AX phosphorylation along the chromatin [reviewed in (18)]. ATM further phosphorylates numerous downstream substrates including the checkpoint kinase 2 (CHK2) (19), which activates the transcription factor p53 by phosphorylation (20). Beside the key role of p53 regulating cellular processes such as apoptosis, senescence and DNA repair [reviewed in (21)], the p53 also regulates the cell cycle progression via accumulation of the cyclin-dependent kinase inhibitor p21. Accumulated p21 inhibits the cyclinE/CDK2 and thereby blocking the G_1/S transition (22). Prolonged activation of p21 after DNA damage is associated with a terminal proliferation arrest, e.g. senescence. In the absence of p53, damaged cells are not able to arrest in the G_1/S phase (23) (Fig.2).

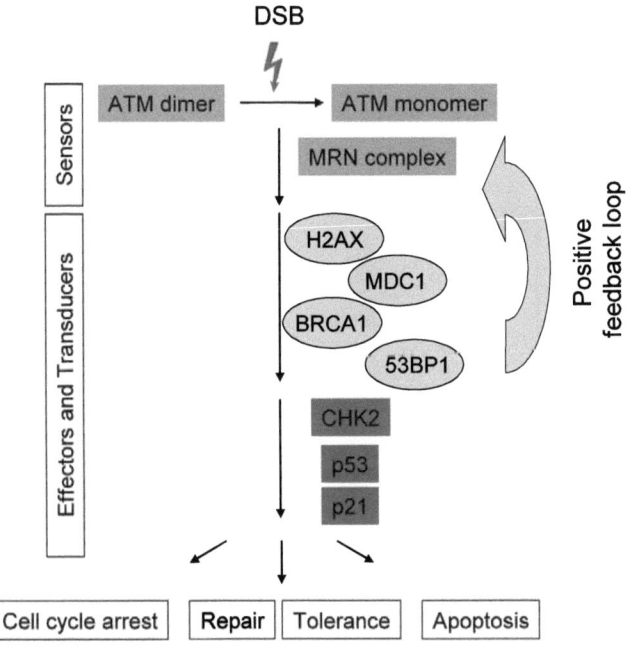

Fig.2 Schematic diagram of the DDR cascade activated by the induction of DNA double strand breaks. Blue boxes refer to sensors, light blue boxes to transducers and red boxes to effectors. For details see text.

Replication fork blockage can result in the generation of replication protein A (RP-A) coated ssDNA leading to the recruitment of the heterodimeric complex (ATR) with its DNA binding subunit ATR interacting protein (ATRIP) (24). The recruitment of ATR-ATRIP at the site of the DNA lesion is boosted by the mediators RAD17/ replication factor C_{2-5} (RF-C_{2-5}) and the RAD9-HUS1-RAD1 (9-1-1) complex and causes the local phosphorylation of γH2AX (25). Additionally, ATR activity is stimulated by the DNA topoisomerase-II-binding protein (TOPBP1). Phosphorylation of checkpoint kinase 1 (CHK1) is mediated by ATR (25) and needs the stimulation by claspin [reviewed in (26)]. Activated CHK1 triggers ubiquitin- and proteasome-dependent degradation of cell division cycle 25 (Cdc25) phosphatases, which subsequently leads to a cell cycle arrest (Fig.3).

The induction of a cell cycle arrest allows the cell to induce repair and/or tolerance mechanisms to maintain genomic stability thereby preventing tumorigenesis [(27, 28) and reviewed in (29)].

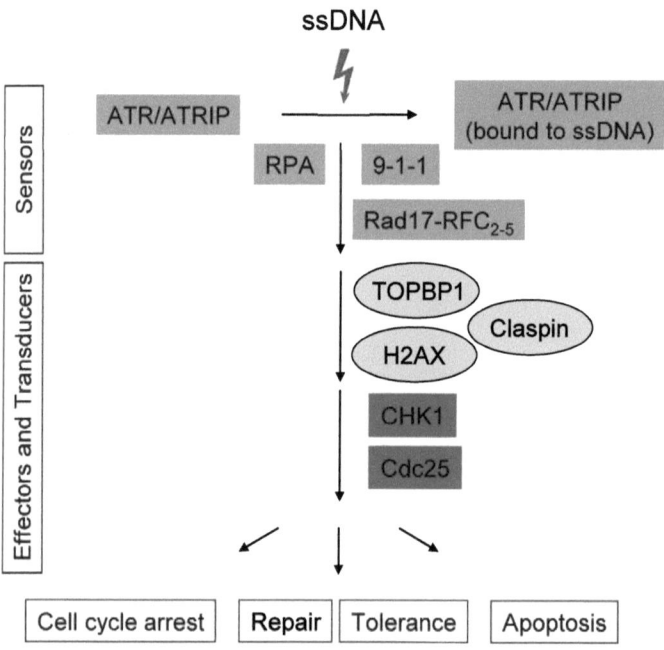

Fig.3 Schematic diagram of the DDR cascade activated by the presence of single stranded DNA (ssDNA), which can be generated after DNA replication fork blockage. Blue boxes refer to sensors, light blue boxes to transducers and red boxes to effectors. For details see text.

The cellular DNA repair machinery consists of non-homologous end joining (NHEJ) and homologous recombination (HR) to repair double strand breaks, base excision repair (BER) to counteract modification of the nitrogenous bases, nucleotide excision repair (NER) to excise altered nucleotides such as 6-4 photoproducts and cyclobutane pyrimidine dimers (CPDs), mismatch repair (MMR) to exchange mispaired nucleotides and direct damage repair for reversal of alkylated nucleotides (Fig.1) [reviewed in (30)]. Although DNA repair processes are not as accurate as high-fidelity DNA replication, DNA repair is considered to be error-free compared to DNA damage tolerance mechanisms.

Activation of the DNA damage tolerance mechanisms is mediated by modifications of proliferating cell nuclear antigen (PCNA). Monoubiquitinated PCNA activates the bypass of the DNA lesions by TLS Pols whereas the polyubiquitination of PCNA by RAD5 triggers the error-free DNA damage tolerance carried out by template switching including fork reversal or recombination past the lesion [reviewed in (31, 32)]. TLS Pols can generate mutations by the incorporation of inaccurate nucleotides according to Watson-Crick base pairing thereby allowing DNA replication to bypass DNA lesions, which otherwise would block replication. The presence of error-prone TLS Pols reflects a trade-off between the maintenance of genomic integrity by avoiding replication fork breakdown and subsequent chromosomal instability and the occurrence of mutations on the nucleotide level by the DNA damage bypass reaction mediated by the TLS Pols. In addition, the low fidelity of TLS contributes to other essential cellular mechanisms, e.g. the TLS Pol ζ contributes directly to class switch recombination (CSR) and indirectly to somatic hypermutation (SHM) in the diversification process of immunoglobulin (IgG) (33).

Alternatively, if the amount of DNA lesions are above a certain threshold or if the type of DNA lesion are irreparable for accurate repair or bypass, cells undergoes apoptosis or senescence due to constitutively activated DDR signaling thereby avoiding chromosomal instability and chromosomal aberrations that could lead to diseases including cancer. It remains still unclear what exactly determines the induction of senescence or apoptosis. It is discussed that the cell fate after DNA damage induction is not only dependent on the duration and nature of the DNA damage but is also dependent on the cell type and on extracellular signals. It was shown that cells undergo senescence due to the induction of severe or irreparable DNA damage that leads to the accumulation of persistent DNA damage (34). The induction of senescence leads to an irreversible growth arrest of the cell and to the development of a complex **s**enescence-**a**ssociated **s**ecretory **p**henotype (SASP) that includes the secretion of cytokines such as IL-6. The initiation and maintenance of SASP needs the DDR factors ATM, NBS1 and CHK2 (34). It has to be further elucidated whether the activation of the SASP has a beneficial or detrimental effect on tumor growth. However, senescent cancer cells are unable to further proliferate and therefore the induction of senescence mirrors a break in tumor progression.

4. Mammalian TLS Polymerases: state-of the art

4.1. History and Discovery

The first genes encoding for TLS Pol *REV1* (encoding the TLS Pol Rev1) and *REV3* (encoding the catalytic subunit of TLS Pol ζ) were discovered and identified in *Saccharomyces cerevisiae* (S. *cerevisiae*) using a screen for reversionless (rev) mutants unable to revert an auxotropic marker after UV irradiation (35). A similar strategy lead to the discovery of UmuC in *Escherichia coli* (36) and REV7, the structural subunit of TLS Pol ζ, in S. *cerevisiae* (37).

Other TLS Pols such as Pol η (eta; *hRAD30A/XPV*), Pol ι (iota; *hRAD30B*), Pol κ (kappa; *DINB1*) and Pol θ (theta; *POLQ*) were identified by searches for homologues of genes of previously identified TLS Pols such as *REV1*, *umuC*, *dinB* and *mus308* (38-41). The TLS Pol μ (mu) (42, 43) and Pol λ (lambda) were discovered and described more recently (42, 44). The most recently described TLS Pol ν (nu) was found due to homology with *mus308* (45).

The ability of TLS Pols to bypass DNA lesions, was firstly described for the yeast TLS Pol ζ (46). Together with yeast REV7, the structural subunit of the TLS Pol ζ (37), yeast REV3 mediates the bypass of Thymine-thymine cyclobutane pyrimidine dimers (TT-CPDs) (46).The property of Rev1 to inserts deoxycytidine monophosphate (dCMPs) opposite abasic sites was first described in yeast (47). Subsequently, the role of human TLS Pol ζ to bypass DNA lesions (48) and the function of human Rev1 as dCMP transferase opposite abasic sites (49) were proposed. The error free TT-CPD bypass activity of human TLS Pol η was discovered by the fact that xeroderma pigmentosum variant (XPV) patients show increased susceptibility to UV-induced skin cancer [reviewed in (50)] and hypermutability (51) due to a defect of TLS Pol η. The human TLS Pol ι was shown to be able to incorporate deoxynucleotides opposite the 3' T of (6-4) photoproducts and abasic sites (52) and opposite N2-adducted guanine (53). The human TLS Pol κ protects cells against the lethal and mutagenic effects of benzo[a]pyrene incorporating the accurate nucleotide opposite the DNA adduct (54). The human TLS Pol θ was shown to be implicated in TLS synthesis and somatic hypermutation (SHM) (55-57). Moreover, the TLS Pol λ and TLS Pol μ are both implicated in V(d)J recombination (58, 59). Further, it was shown that TLS Pol ν is able to bypass thymine glycols (60).

4.2. Fidelity and activation of TLS

TLS Pols are optimized to bypass DNA lesions and are characterized by low fidelity DNA synthesis and lack the exonuclease proofreading activity thus generating mutations. The error rate of DNA replication Pols of the families A, B and C including correct incorporation of the nucleotide and the proofreading activity is between 10^{-6} and 10^{-8}. Auxiliary proteins such as PCNA and RP-A (61) and postreplicative MMR decrease the error rate to 10^{-8} and 10^{-10}. The error rate of the TLS Pols ranges from 10^{-1} to 10^{-3} for replication of undamaged DNA [reviewed in (62, 63)].

Although the location in the tertiary structure consisting of palm, thumb and fingers is conserved among the different Pol families, the thumb and fingers are smaller in the TLS Pols (Fig.4). Compared to the DNA replication Pols where the fingers tightly bind the incoming dNTPs and make a conformational change upon correct Watson-Crick base pairing, the active site of TLS Pols is more open and less constrained to reject wrong paired base pairs. Therefore, TLS Pols are able to mediate the bypass reaction of non-coding DNA lesions. The additional little finger of the Y family TLS Pols supports the stabilization of the template DNA and influences fidelity and activity (64) (Fig.4).

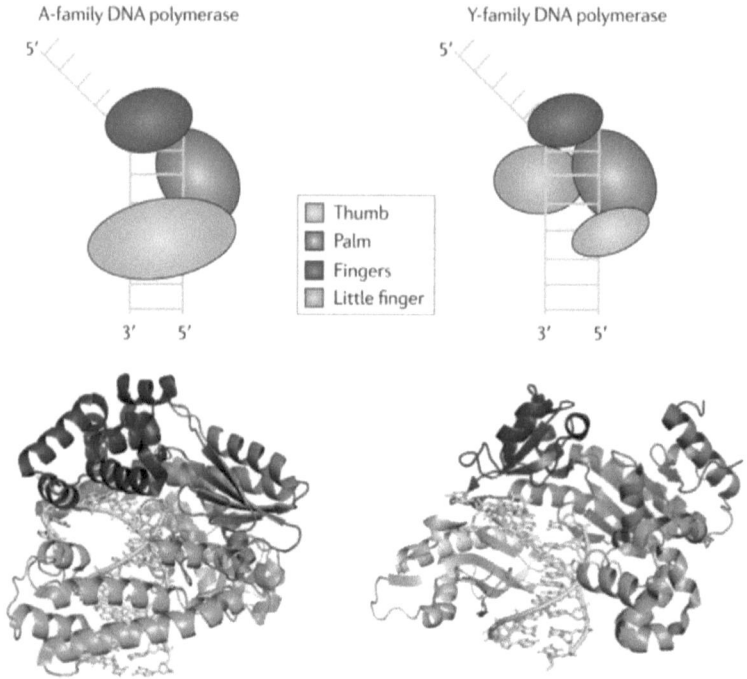

Fig.4 Structural comparison of the *Thermus aquaticus* DNA Pol I of the A-family and TLS Pol η of the Y-family [reproduced from ref. (65)]. For details see text.

A key event in the activation of TLS is the monoubiquitination of PNCA that encircles the DNA and achieves the TLS processivity. This event is proposed as a switch between DNA replication and TLS based on observations of the Y-family TLS Pols whereas TLS Pols bind to monoubiquitinated PCNA through the ubiquitin-binding domains such as the ubiquitin binding motif (UBM), the ubiquitin binding zinc finger (UBZ)] and a PCNA interacting peptide box (PIP) (66, 67). The monoubiquitination of PCNA is carried out by the RAD6-RAD18 complex upon replication fork stalling [reviewed in (31, 32)]. Moreover, a role of p53 via the PCNA-interacting domain of p21 has been proposed to mediate the activity of TLS Pols, together with the DNA damage induced monoubiquitination of PCNA at lysine 164 (68).
An interplaying mechanism was proposed between the deubiquitinating enzyme USP1 and ubiquitination factors including RAD6 and RAD18. USP1 acts as a negative regulator, thus

removing the ubiquitin residue from the monoubiquitinated PCNA to reduce the mutagenic effect by TLS Pols [reviewed in (69)].

In addition to the discussed activation of either the error-free or the error-prone tolerance mechanisms upon DNA lesions, the question raised whether the regulation between DNA replication and the DNA damage tolerance (DDT) pathway is dependent on the cell cycle. It has been shown, that inappropriate expression of TLS Pol κ and Pol β in human cells leads to accumulation of mutations by competing with the high fidelity replicative DNA Pols. This finding indicated that the activation of high-fidelity replicative Pols and TLS Pols has to be temporally regulated (70, 71)

A model was proposed where the major function of TLS Pols is to allow replication to continue in the presence of DNA damage thereby allowing progression in S-phase (72). However, initial studies in yeast revealed that TLS Pols have a function beyond S-phase, the so called post replicative repair (PRR) (73).

Moreover, other studies in yeast revealed that TLS Pols are also required for ICL repair during the G_1 phase (74).

Further, REV1 and PCNA, which regulate TLS activity, are also active outside S-phase. In detail, REV1, which interacts and thereby regulates the activity of several TLS Pols is highly expressed in late S and early G_2 in yeast (75). Ubiquitinated PCNA in human cells is stably bound to chromatin even after the lesion has been removed (76). Recently, a distinct role for REV1 and PCNA in the regulation of TLS was discussed using the chicken cell line DT40. It was shown, that REV1 plays a role for maintaining replication fork progression at damaged DNA and PCNA is required for postreplicative action (77). Additionally, studies in mammalian revealed that human REV3 is accumulated in G_1 and at the G_2/M transition in U2OS cells (78). However, recent findings in yeast revealed, that both error-prone TLS and error-free template switching takes place in G_2/M phase after DNA replication (79).

Taken together, the activation of the DDT pathway is regulated by the mono- or polyubiquitination of PCNA and the activation of REV1, which itself might be dependent on the stage of the cell cycle. More studies are needed to elucidate the exact details of the regulation of the DDT pathway.

4.3. TLS Polymerase families

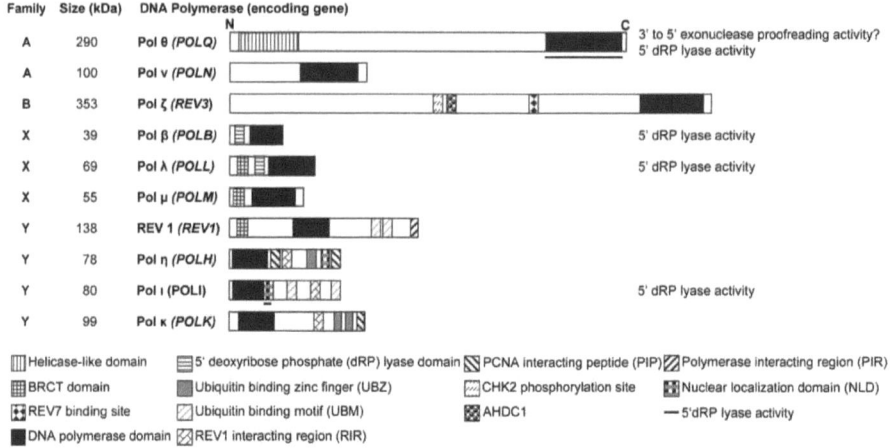

Fig.5 Overview of TLS Pols [Adapted from ref. (65)]. For details see chapter 4.3.

4.3.1. Family A: TLS Polymerases theta (θ) and nu (v)

4.3.1.1. TLS Pol θ

The A family Pols consist of DNA Pol γ, TLS Pol v and TLS Pol θ. TLS Pol θ was mapped on the chromosome locus 3q. The cDNA of TLS Pol θ shows the typical Pol motifs A, B and C for A family Pols at the C-terminal region and an ATP domain at the N-terminal region. The *POLQ* gene encodes a protein of 2592 amino acids (aa). The protein sequence shares homology to the Mus308 protein of Drosphila melanogaster (41, 80). Further research estimated a protein of 290 kDa, with an N-terminal ATPase helicase domain and a C-terminal Pol domain. Despite the ATPase helicase domain, no helicase activity could be detected so far (80). It was shown that TLS Pol θ is able to bypass apurinic/apyrimidinic (AP) site and thymine glycol (Tg). TLS Pol θ preferentially incorporates adenine opposite the AP site and efficiently achieves the subsequent proceeding of the replication using the incorporated nucleotides as a primer (henceforth referred to as extension step) (81). Moreover, it can carry out the extension step from mismatches after error prone dNTP incorporation by human TLS Pol ι (82) or *S. cerevisiae* TLS Pol ζ opposite (6-4) photoproducts *in vitro* (52).The fidelity of TLS Pol θ during dNTP incorporation is lower than usual for A family Pols (81), since it lacks the 3' to 5' exonuclease proofreading activity (80). It is proposed that TLS Pol θ has a function in SHM of immunoglobulin diversification by misincorporation over AP sites and low fidelity DNA synthesis (55-57). In contrast, it was suggested that human TLS Pol θ from HeLa cells nuclear extracts synthesizes DNA with a high fidelity and possesses 3' to 5' exonuclease proofreading activity (83).

TLS Pol θ mutant mice show an increase of spontaneous and radiation-induced micronuclei suggesting a role for TLS Pol θ to maintain chromosomal stability (84, 85) and TLS Pol θ knockout chicken DT40 B-cell line shows hypersensitivity to hydrogen peroxide (H_2O_2). Furthermore, CH12 mouse B lymphoma cells with a knockdown of TLS Pol θ showed elevated sensitivity to UV irradiation, mitomycin C (MMC), cisplatin, etoposide, Ionizing irradiation (IR) and methyl methanesulphonate (MMS) (86). Additionally, TLS Pol θ has been shown to have 5' dRP lyase activity that is involved in short patch BER *in vitro* (87).

4.3.1.2. TLS Pol v

The full length *POLN* gene comprises 24 exons with a length of 900 aa and is located on chromosome 4p16.2 that is deleted in approximately 50% of breast carcinomas (88). The *POLN* gene encodes a protein with a size of 160 kDa. The C-terminal Pol domain of TLS Pol v consists of the typical A family Pol motifs A, B and C and shares 29% identity with the C-terminus of TLS Pol θ. Neither a 3' to 5' nor a 5' to 3' nuclease domain were identified (45).

In vitro experiments showed the ability of TLS Pol v to bypass Tg (60). Interestingly, it has been shown that TLS Pol v is involved in homologous recombination and cross link repair. Downregulation of *POLN* expression reduces HR and sensitizes HeLa cells to DSB inducing agents such as campothecin. Moreover, TLS Pol v interacts with factors of the Fanconia anemia pathway that are involved in crosslink repair and inhibition of *POLN* expression sensitizes HEK293T cells to the DNA crosslinking agent MMC.

TLS Pol v interacts with the helicase HEL308 during DNA repair which shares homology with the *POLN* in the *Mus308* gene of D. melanogaster (89).

Surprisingly, DT40 *POLN* knockout cells are not sensitive to crosslinking agents such as cisplatin, MMC and the DSB inducing agent campothecin (90, 91).

4.3.2. Family B: TLS Polymerase zeta (ζ)

4.3.2.1. TLS Pol ζ

The family B includes the highly accurate DNA Pols δ (delta), ε (epsilon), α (alpha), and the error-prone TLS Pol ζ (92, 93). Unlike the DNA Pols δ and ε, TLS Pol ζ lacks the 3' to

5' exonuclease proofreading activity. The human TLS Pol ζ and its yeast homologue are heterodimeric proteins consisting of the catalytic subunit REV3 and the structural subunit REV7 (46, 94).

The human REV3 protein has two transcripts that have a length of 3052 and 3130 aa and the larger protein has a size of 353 kDa compared to 173 kDa of the yeast REV3. The big differences between the yeast REV3 and the human REV3 is due to the exon 13 with a length of 1388 aa. Human REV3 shows ~36% identity with the N-terminal region, ~29% identity with the central REV7 binding region and ~39% identity with the C-terminal DNA Pol region. The C-terminal region consists of six B-family conserved DNA Pol motifs and two zinc finger motifs (95-97) (Fig.6). Human *REV3* is located on chromosome 6q21 and its mouse equivalent on chromosome 10 (98, 99). Interestingly, the location of *REV3* on the chromosome 6q21 is within the fragile site FRA6F, which is known to be commonly deleted in several types of human leukemias and solid tumors (100). The human *REV3* consists of an out-of-frame ATG in the 5' region that reduces the rate of correct transcripts. Moreover, a sequence upstream of the AUG initiator codon has the potential to form a stem-loop hairpin that lowers the rate of translation. It is suggested, that together with the alternative splicing form, this features of the primary and secondary structure leads to a low expression level of *Rev3* (96, 98). Indeed, the protein concentration of REV3 in xenopus laevis egg extracts is much lower than those of other replication and repair proteins and does not change within the early embryonic development (101). Despite the human REV3 contains a REV7 binding region, so far, no interaction between full-length REV3 and REV7 could be demonstrated. However, it was shown, that a human REV3 fragment interacts with full-length REV7 and a part of human REV7 interacts with human REV3 and REV1 (102). Moreover, within the exon 13 of REV3, serine 995 was shown to be phosphorylated by CHK2 (78). Additionally, a telomere repeat factor homology (TRFH) docking motif site, homologies to an AT hook DNA-binding motif containing protein1 (AHDC1) and a predicted KIA2022 protein were located within the exon 13 (Fig.6).

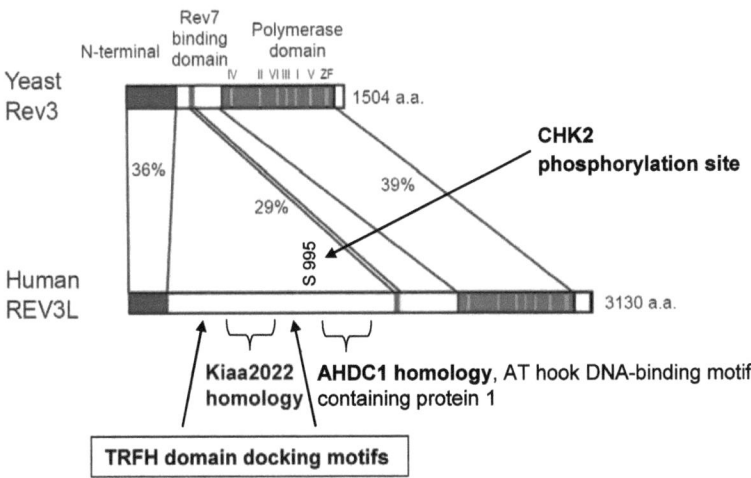

Fig.6 Comparison between the yeast REV3 and the human REV3 [adapted from (103)]. For details see text.

The human TLS Pol ζ is thought to be the major error-prone TLS Pol bypassing DNA lesions. In contrast to viable *Rev3* null yeast mutants, the disruption of *Rev3* in mice causes embryonic lethality around midgestation (104-107). It is known that during the early stages of embryogenesis checkpoints are actively silenced (108) to allow rapid cell division and it was proposed that REV3 is essential during this strict temporal program. The embryonic lethal effect could not be rescued by the absence of p53 suggesting a p53-independent pathway. However, mouse embryonic fibroblasts (MEFs) with a p53-deficient background could be generated (109, 110).

An *in vitro* experiment showed a direct regulation of *Rev3* expression level through a p53 responding element in the *Rev3* promoter region. In addition, *Rev3* expression was increased after DNA damage induction in a p53-dependent manner (111). An independent study also showed increased *Rev3* mRNA level after cisplatin treatment (112).

REV3 downregulation in human foreskin fibroblasts revealed decreased mutation frequency after treatment with UV or benzo[α]pyrene diolepoxide (113). Similarly, MEFs derived from mice expressing *Rev3* antisense revealed decreased mutagenic frequency after UV treatment (114). It was shown that *Rev3* knockout in a p53-deficient background in MEFs leads to increased chromosomal instability (109). Moreover, in mice with a conditional deletion of REV3, thymic lymphomas occurred with decreased latency and elevated incident in a p53-deficient background (115). Recent *in vitro* and *in vivo* studies revealed that *Rev3* inhibition increased the sensitivity of lymphoma to cisplatin (116). Similar, a non small cell lung cancer cell line (NSCLC) with downregulated REV3 transplanted in p53 deficient mice showed a decreased growth rate and prolonged survival upon cisplatin treatment. Additionally, an *in vitro* mutagenesis assay showed a reduced frequency of 6-thioguanine resistant colonies after an initial treatment with cisplatin in REV3 deficient cells compared to the control (116, 117). These findings assume a role of REV3 in the formation of resistance due to induction of mutations.

Overexpression of *REV3* in yeast led to an elevated rate of UV-induced mutagenesis (118). This finding together with the embryonic lethal effect after REV3 abrogation indicates that the level of REV3 has to be tightly regulated to maintain genomic integrity. Recent findings propose TLS Pol ζ to have a function not only in TLS synthesis but also in DNA repair. *Rev3* disrupted DT40 chicken cells shows increased sensitivity to various DNA damaging agents including UV, methylmethane sulphonate (MMS), cisplatin and ionizing radiation (IR) (119). XPV cells treated with siRNA against TLS Pol ζ reveal increased UV sensitivity (120). *Rev3* deletion impairs specific DNA repair mechanisms as homologous recombination (HR) (112) and ICL repair (121, 122). Furthermore, somatic hypermutation (SHM) and/or class switch recombination (CSR) of immunoglobulin (IgG) were affected by TLS Pol ζ ablation (33). In addition, a possible role of REV3 in the telomere maintenance can be suggested, due to the co-purification of REV3 with TRF2 (123) and the presence of a predicted TRFH docking motif site within the exon 13 (http://elm.eu.org/). REV3 shares the AT-hook domain with AHDC1, which has been proposed to be phosphorylated by either ATM and ATR upon DNA damage, indicating a putative regulation of REV3 by ATM and ATR (124). It is hypothesized that REV3 could be regulated by DDR, since REV3 is phosphorylated by either CHK2 and ATM and ATR. The KIA2022 gene, which shares a conserved region with REV3, is proposed to be absent in patients with severe mental retardation and is highly expressed in fetal brain and in the adult cerebral cortex, suggesting a role in brain development and/or cognitive function (125). In this context, murine REV3 was first identified as a gene induced by treatment of primary cultured cerebral cortical cells with the seizure-inducing agent pentylentetrazol (126).

The human REV7 protein has a length of 211 aa and a size of 24 kDa and shares ~23% identity with the yeast REV7. The human REV7 is located on the human chromosome 1p36. Moreover, it displays ~23% identity with the spindle checkpoint assembly protein MAD2. Therefore, REV7 is also known as MAD2B and MAD2L2 in higher eukaryotes. The REV7 contains a HORMA (Hop1/Rev7/Mad2) domain that is known to interact with chromatin (127). Additionally, REV7 interacts with CDH1 and CDC20 of the anaphase-promoting complex/ cyclosome (APC/C) (128) and the protein MAD2, a spindle checkpoint protein (94), indicating that REV7 is involved in the regulation of mitosis. Interestingly, the bacterial pathogen *Shigella* delivers IpaB effectors into epithelial cells to efficiently colonize the epithelium. It has been shown that IpaB interacts with MAD2L and leads to cell cycle arrest (129).

4.3.3. Family X: DNA Polymerase beta (β) and TLS Polymerases lambda (λ) and mu (μ)

The Pols of the X family include DNA Pol β, terminal deoxynucleotidyl transferase (TdT), TLS Pol λ and TLS Pol μ. All the X family Pols lack the 3' to 5' exonuclease proofreading activity.

4.3.3.1. DNA Pol β

DNA Pol β is a 39 kDa monomeric protein and the encoding gene *POLB* is located on chromosome 8 in both mice and human (130). DNA Pol β consists of two protease resistant segments linked by a short protease sensitive segment indicating that DNA Pol β activity might be controlled by proteolytic activity. The 8 kDa N-terminal lyase domain shows a strong affinity to ssDNA (131), whereas the 31 kDa C-terminal Pol domain specifically binds double-stranded nucleic acids (132). The 31 kDa Pol domain consists of three subdomains. The catalytic subdomain (palm) coordinates two metal-ions and mediates the nucleotidyltransferase reaction (133) and the other subdomains mediate the binding of duplex DNA (thumb) and nascent base pairs (fingers) (134). It has been shown that DNA Pol β is able to fill short gaps in double stranded DNA (135). The 8 kDa lyase domain was shown to direct DNA Pol β to phosphorylated 5' side of a DNA gap for its bypass (136) and recently to mediate the removal of 5' dRP from the AP site via β-elimination after the incision step by the AP endonuclease (137). Beside the role of DNA Pol β in short patch BER, a function in long patch BER was proposed. Both short- and long-patch repair are impaired after DNA Pol β ablation (138-140). Additionally, DNA Pol β has a role in bypassing DNA lesions such as cisplatin-DNA adducts (141). It was shown that DNA Pol β is not involved in the diversification of IgG (142). DNA Pol β knockout mice are not viable reflecting the important role of Pol β during embryonic development (143).

NIH 3T3 mouse fibroblasts with downregulated DNA Pol β showed increased drug sensitivity to cisplatin and UV-induced DNA lesions, which are repaired by NER, suggesting a role of DNA Pol β in NER (144). Further, DNA Pol β protects MEFs against the cytotoxicity of oxidative DNA damage (145) and the mutagenic effect of methylating agents such as MMS (146). Similarly, fibroblasts from DNA Pol β null mice were hypersensitive to the methylating agent MMS due to the missing lyase activity of DNA Pol β (147).

In addition, ectopic expression of DNA Pol β leads to aneuploidy, aberrant localization of the centrosome-localized γ-tubulin protein during mitosis, checkpoints defects *in vitro* and tumour induction *in vivo* (148). Moreover, DNA Pol β is upregulated in chronic myelogenous leukemia (CML) patients (149). These results indicate that a tight regulation of DNA Pol β is essential to maintain genomic integrity.

4.3.3.2. TLS Pol λ

TLS Pol λ has a size of approximately 69 kDa and its gene *POLL* is located on the chromosome 10 in human and on the chromosome 19 in mice (42, 44). The human TLS Pol λ consists of 575 aa and shares 32% residue identity to Pol β consisting of the C-terminal Pol domain including palm, thumb and fingers and the 8 kDa 5' dRP lyase domain. Additionally, TLS Pol λ contains an N-terminal BRCT (BRCA1 C-terminus) domain followed by a serine/proline rich region that is absent in Pol β (150). The BRCT domain is widely found in proteins involved in DNA damage repair and DNA damage checkpoint control (151). TLS Pol λ shows terminal deoxynucleotidyl transferase activity (150), which prefers the incorporation of pyrimidine nucleotides (152). The TLS Pol λ has been shown to be less accurate for base substitutions and much less accurate for single-base deletions (153). Further, TLS Pol λ is unable to differentiate between matched and mismatched primer termini during the extension step, therefore suggesting TLS Pol λ as a candidate for NHEJ and as mismatch extender during TLS (153, 154). Additional *in vitro* studies showed that TLS Pol λ require their BRCT domain and is physically and functionally dependent on Ku during NHEJ (155). Recently, it has been shown, that a TLS Pol λ variant consisting of a single nucleotide polymorphism (SNP), a cytosine/thymine variation, leads to increased mutation frequency, chromosomal aberration and defects in NHEJ (156).

Moreover, TLS Pol λ is also discussed to participate in BER. Uracil-containing DNA was efficiently repaired in an *in vitro* reconstituted BER reaction by the 5' dRP lyase activity of

TLS Pol λ, in coordination with its polymerization activity (157). TLS Pol λ null mice are viable and fertile, but shortening of the heavy chain coding joints was reported (58).

4.3.3.3. TLS Pol μ

TLS Pol μ has a size of 55 kDa and its gene *POLM* is located on the human chromosome 7. It consists of 492 aa and shares 42% identity to TdT (42, 44). TLS Pol μ, as TLS Pol λ, contains a Pol domain and a BRTC domain. In contrast to Pol β and λ, the TLS Pol μ lacks a 5' dRP lyase activity (157). Isolated TLS Pol μ is highly error-prone for frameshifts during DNA synthesis. Interestingly, TLS Pol μ is able to extend from mismatches by frameshift synthesis mechanism and thereby promoting microhomology search and microhomology pairing between the primer and the DNA template (158). Moreover, TLS Pol μ shows template-independent Pol activity under physiological conditions (Mg 2^+ present) preferring the incorporation of pyrimidines and thereby generating terminal microhomology, which can be ligated by the XRCC4:DNA ligase IV (159) All these findings suggesting TLS Pol μ as a candidate for NHEJ of DSBs. It has been shown that TLS Pol μ, as TLS Pol λ, interact with the Ku-DNA complex through its BRCT domain (155)

Interestingly, TLS Pol u null mice are viable and fertile, but they show impaired V(D)J recombination due to shortening of the light chain coding ends, but not of the heavy chain coding ends (58, 59)

4.3.4. Family Y: Rev1, TLS Polymerases eta (η), kappa (κ) and iota (ι)

The human TLS Pol members of the Y family include REV1, Pol η, Pol κ and Pol ι [reviewed in (160)]. All the Y family members lack the 3' to 5' exonuclease proofreading activity [reviewed in (63)] and share a general conserved N-terminal Pol domain for the catalytic activity and a non-conserved C-terminus, which, at least for the human TLS Pol ι (161) and Pol κ (162), is responsible for the regulation of the activity. The conserved N-terminus of the Pol domain includes five motifs (I to V) corresponding to the catalytic core complex of the Pol. Motif I and II form the catalytic epicentre (palm) with its three acidic residues harbouring the two metal ions mediating the nucleotide transfer. Despite sequence differences the palm domain with its nucleotide transfer function is widely conserved between Y-family TLS- and A- and B-family replicative DNA Pols. The motif III and IV belong to the finger and thumb domain, respectively. They bind the triphosphate of the nascent incoming dNTP and mediate the incorporation of the nascent nucleotide whereas the additional motif V binds the primer strand. The C-terminus of the motif V resides either the so called little finger (LF), polymerase associated domain (PAD) or the wrist that supports the DNA synthesis activity and is conserved and unique among the Y-family TLS Pols (64, 163, 164) (Fig.4).

4.3.4.1. Rev1

The 1251 aa human REV1 protein has a M_r of 138 kDa and is encoded by the gene REV1, located on chromosome 2. Beneath the typical Y-family conserved domains, a BRCT domain that binds phosphorylated proteins and thereby mediating cell cycle control upon DNA damage is located at the N-terminus (151). At the C-terminal end are two UBMs (66) followed by a TLS Pol η, Pol ι and Pol κ interaction region (165).
It has been proposed that the nucleotide insertion activity is not the main function of REV1 but that REV1 helps to coordinate the Pol switch between the normal- and the substituting TLS Pol upon PCNA monoubiquitination. Although it was shown that murine REV1 binds monoubiquitinated PCNA via its UBMs, it is assumed that the Pol switch function of REV1 might also be dependent on the other protein interaction domains of REV1, e.g. the BRCT and the Pol interaction region. In detail, the C-terminal end of human Rev1 is able to interact with several TLS Pols including Pol η, Pol ι, Pol κ and Pol ζ, thus supporting the assumption that Rev1 acts as a scaffold protein for several TLS Pols (102, 165, 166).
Murine Rev1 binds ubiquitin through its UBMs and thereby mediating its localization to DNA damage foci. UBM mutants showed increased mutational aberrations after UV irradiation and elevated sensitivity to UV irradiation and cisplatin. This effect of sensitization was even increased in UBM and BRCT double-mutants (67).
Additionally, murine REV1 binds PCNA through its BRCT domain and the monoubiquitination of PCNA enhances this reaction (167). Rev1 ablation sensitizes DT40 chicken cells to various DNA damaging agents including cisplatin, UV irradiation and MMS. Additionally, Rev1 is required for the maintenance of chromosomal stability after UV irradiation (167). An *in vivo* mouse model showed that REV1 inhibition in B-cell lymphoma reduces cisplatin– and cyclophosphamide (CTX) induced mutagenesis and prolongs survival upon CTX treatments (116).
Recently, it has been shown *in vitro* that Rev1 silencing impairs its replicating over G-quadruplex (G4) structures and subsequent biased incorporation of newly synthesized histones due to failure to use of recycled histones, resulting in changes of the epigenetic

pattern. It is suggested that the ablation of Rev1 and therefore the inaccurate recycling of histones lead to a decrease of chromatin dependent gene silencing (168).

4.3.4.2. TLS Pol η

Human TLS Pol η consists of 713 aa and is encoded by the *POLH* (Xeroderma pigmentosum variant, XPV) gene, localized on chromosome 6. TLS Pol η has a size of 78 kDa. Additional to the N-terminal conserved Pol domain, TLS Pol η consists of a Rev1-interacting region (RIR), an ubiquitin binding zinc finger (UBZ), a nuclear localization domain and a PCNA interaction peptide box (PIP). XPV cells are sensitive to UV irradiation and show an increased mutagenic rate despite functional NER indicating that TLS Pol η bypasses the UV lesions in a non-mutagenic manner (169). It is proposed that in the absence of TLS Pol η, TLS Pol ι serves as the error prone Pol which bypasses the UV induced lesions (170). Also Rev1 is suggested to have a regulatory role in TLS of UV induced lesions (171). TLS Pol η needs for its TLS activity the PIP and the UBZ domain for binding the monoubiquitinated PCNA. Mutation in either the PIP or the UBZ domain increases the UV sensitivity (66).

Interestingly, it has been shown that loss of TLS Pol η in mice leads to decrease in adenine/thymine mutations during SHM of IgG indicating that TLS Pol η bypass adenine and thymine in an error prone manner (172).

4.3.4.3. TLS Pol κ

The human TLS Pol κ is encoded by *POLK* gene and has a length of 870 aa. The TLS Pol κ has a size of 99 kDa and is located on chromosome 5. The N-terminal part consists of the conserved Pol domain whereas the variable C-terminus consists of a RIR, two UBZ and a PIP. It has been shown that TLS Pol κ co-localizes to a lesser extend with PCNA at replication foci after UV irradiation, hydroxyurea or benzo[α]pyrene treatment compared to TLS Pol η (173). It has been shown that ES cells deficient in the TLS Pol κ gene are more sensitive and acquire more mutations after treatment with benzo[α]pyrene and that TLS Pol κ bypasses BP-G accurately and efficiently *in vivo* (54). Additionally, XPV cells treated with siRNA against TLS Pol κ reveal increased UV sensitivity (120) indicating that TLS Pol κ is able to bypass UV induced DNA lesions. Recent findings suggest that TLS Pol κ has a function during NER and that its activity is dependent on RAD18 and monoubiquitinated PCNA (174).

The role of TLS Pol κ in promoting tumorigenesis has been discussed since ectopic expression of Pol κ leads to DSBs, aneuploidy and tumorigenesis in nude mice (71).

4.3.4.4. TLS Pol ι

Human TLS Pol ι is encoded by the *POLI* gene consists of 715 aa. The TLS Pol ι has a size of 80 kDa and is localized on chromosome 18. TLS Pol ι shares with the other Y family members the N terminal conserved Pol domain and at the variable C-terminal a RIR, two UBMs and the PIP. In contrast to TLS Pol η and Pol κ the PIP of TLS Pol ι is located in the middle of the protein between the conserved Pol domain and the Rev1 binding domain. Interestingly, TLS Pol ι possess a 5' dRP lyase activity (175) that is located within a nuclear localization domain (176).
As in the case of TLS Pol η, the PIP and the UBM domains of TLS Pol ι are important for localization to the replication fork by binding monoubiquitinated PCNA (66). The localization and accumulation of TLS Pol ι to stalled replication forks is widely dependent on TLS Pol η through physical interaction (177). Recently, it has been shown that human fibroblasts in which TLS Pol ι is stably downregulated exhibits increased sensitivity against H_2O_2 and menadione and BER is decreased (176). Additionally, after H_2O_2 treatment, TLS Pol ι binds to chromatin and interacts with the BER factor XRCC1, suggesting TLS Pol ι as repair Pol of oxidative damage (176).

4.4. Translesion synthesis

4.4.1. One- and two-Polymerase mechanism

There are TLS Pols that are able to replicate over a DNA lesion by both incorporating nucleotides opposite the damaged DNA and by extending from the inserted nucleotides. The so called one-Pol mechanism has been shown to be performed by TLS Pol κ replicating over AP sites *in vitro* and by TLS Pol η bypassing UV induced CPDs *in vivo* (Fig.7).

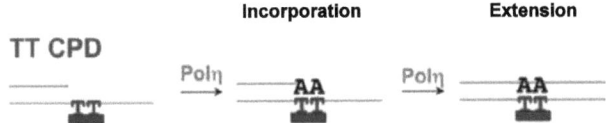

Fig.7 One Pol error-free bypass of a TT-CPD carried out by TLS Pol η [adapted from ref. (178)]. For details see text.

However, other lesions such as BP-G or cisplatin-DNA adducts can not be replicated by a one-Pol mechanism but requires a continuous progression of two Pols, a so called two-Pol mechanism. The first Pol incorporates the nucleotides opposite the DNA lesion and a second Pol subsequently extends from the inserted nucleotides. Depending on the type of DNA lesion, different pairs of Pols interact together to replicate the DNA lesion resulting in either an error-free or error-prone bypass (Fig.8 and Table.1).

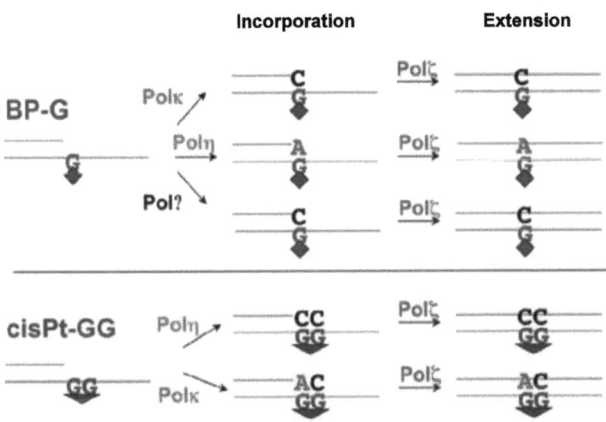

Fig.8 Two-Pol mechanism for bypassing a BP-G or cisplatin-DNA adducts. The first step is achieved by a TLS Pol incorporating a nucleotide in an error-free or error-prone manner subsequently followed by TLS Pol ζ carrying out the extension step [adapted from ref. (178). For details see text.

4.4.2. Apurinic/ apyrimidinic (AP) sites

AP sites present a strong block to continued synthesis by the replicative DNA machinery, Pol β is the primary enzyme used for gap filling DNA synthesis, during BER. However, if AP sites are not repaired, a one-Pol mechanism has been proposed in an *in vitro* assay where the TLS Pol θ preferentially incorporates an adenine opposite the AP site followed by a guanine and cytosine/thymine (81). In addition, it has been shown *in vitro* that TLS Pol η is able to incorporate nucleotides opposite AP sites preferentially adenine and guanine and extend from the incorporated nucleotide favouring an adenine (179). An adenine opposite the AP site is also the best primer for the extension step of TLS Pol θ (81).

In vitro studies showed that isolated TLS Pol λ from calf thymus is able to replicate a damaged (apurinic) DNA template in vitro (180). Similarly, isolated human TLS Pol λ is able to synthesize over an AP site and this bypass is stimulated by PCNA *in vitro* (181) Alternatively, AP sites might also be bypassed by a two-Pol mechanism. It has been suggested that TLS Pol μ is capable to incorporate nucleotides opposite AP sites *in vitro* although in this process deletions are frequently generated due to primer realignment (182). Similary, TLS Pol ι and REV1 are able to incorporate one nucleotide opposite an AP site (183).

Thus, AP sites might also by bypassed by a two-Pol mechanism, with TLS Pol μ, TLS Pol ι or REV1 performing the insertion step followed by second Pol carrying out the extension step.

4.4.3. 7, 8-dihydro-8-oxoguanine (8-oxo-G)

8-oxo-G is generated by oxidative stress and leads to frequent misincorporation (10-75%) of adenine opposite the 8-oxo-G, thus, generating a guanine:cytosine to thymine:adenine transversion. It has been shown that 8-oxo-G lesion *in vitro* can be bypassed by TLS Pol μ resulting in a -1 deletion due to primer realignment during TLS (182). More recently, it has been shown that 8-oxo-G lesion *in vitro* are mainly bypassed by TLS Pol λ and TLS Pol η and that the presence of PCNA and RP-A increases the fidelity of correct cytosine incorporation over the incorrect adenine incorporation opposite the 8-oxo-G by a factor of 1200 fold for TLS Pol λ and by a factor of 68 fold for TLS Pol η (184).

4.4.4. Thymine glycol (Tg)

Tg is the most common thymine lesion induced by reactive oxygen species (ROS) (185). *In vitro* studies have shown, that TLS Pol θ is able to incorporate nucleotides over both 5R- and 5S-diastereoisomers of Tg with similar efficiency but fails to process the extension step (81). Similary, the TLS Pol ν is able to bypass 5S-Tg in an error-free manner whereas the bypass of 5R-Tg was less acurate (60). The TLS Pol λ has been shown to bypass a Tg in gapped DNA structures. Additionally, dependent on the size of the gap, TLS Pol λ is able to possess the extension step. The bypass fidelity of TLS Pol λ is thereby increased by the presence of PCNA (186). Recently, a two-Pol mechanism *in vivo* for error-free Tg bypass including TLS Pol κ as nucleotide inserter and TLS Pol ζ as extender was proposed (187).

4.4.5. (6-4) photoproduct

The major DNA lesion induced by UV irradiation is the (6-4) photoproduct that leads to a 44° bend of the DNA helix and inhibits the DNA synthesis (188, 189). Beside the common repair of (6-4) photoproduct by NER, *in vitro* experiments have shown that TLS Pol μ is able to bypass the (6-4) photoproduct (182). Two alternative two-Pol mechanism models to bypass the (6-4) photoproduct have been proposed using an *in vivo* plasmid based model. In the first model TLS Pol ι and TLS Pol η alternatively incorporate nucleotides at the 3' thymine or 3' cytosine in an error-free or error-prone manner, subsequently extended by a yet unknown DNA Pol. In the second model a yet unknown Pol incorporates nucleotides at the 3' thymine or 3' cytosine in an error-free manner. Subsequent extension is carried out by TLS Pol ζ (190).

4.4.6. Cyclobutane pyrimidine dimer (CPD)

CPDs are DNA lesions generated by UV irradiation and are associated with the hereditary disease XPV due to a lack of TLS Pol η (169). CPDs are replicated *in vitro* by both the one- and the two-Pol mechanism. *In vitro* experiments revealed that TLS Pol μ can bypass a CPD in a mainly error-free manner (182). Further, TLS Pol η is able to replicate efficiently over CPD *in vitro* thereby bypasses the CPD with lower fidelity and higher error rates at the 3' thymine than at the 5' thymine. Additionally, the UV irradiation induced mutagenesis is higher at the 3' base (191). Recently, it has been shown *in vivo* that TLS was reduced and mutagenicity increased in cells lacking TLS Pol η using a quantitative TLS assay system measuring TLS across CPDs. Most of the mutations could be found opposite the 3' thymine of the CPD. Similar to the *in vitro* studies, the error frequency of TLS Pol η *in vivo* is approximately 1% (192).

The TLS across CPD can also be processed by the two-Pol mechanism model. Very recently it has been proposed in an *in vivo* model that CPD in XPV cells, e.g. in the absence of TLS Pol η, is replicated by the two-Pol mechanism in an error-prone manner. The two step model includes either TLS Pol ι, TLS Pol κ or a yet undefined Pol or a combined action for the first and the second pyrimidine nucleotide incorporation opposite a CPD, followed by the extension step achieved by TLS Pol ζ (120).

4.4.7. Benzo[α]pyrene-guanine (BP-G)

BP-G is a major tobacco smoke-induced DNA lesion which is associated with the development of lung cancer (8). *In vitro* and *in vivo* studies showed efficient bypass of BP-G by TLS Pol κ using a gapped plasmid consisting of a BP-G lesion (193).
Additionally, BP-G has been shown to be bypassed by a two-Pol mechanism *in vivo*. The accurate incorporation of the nucleotides is carried out either by TLS Pol η or TLS Pol κ followed by the extension of TLS Pol ζ (194).

4.4.8. Intrastrand-crosslinks

Cross-linking agents including cisplatin, which induces DNA intra- and interstrand crosslinks (ICLs) are widely used in anticancer chemotherapy. Intrastrand-crosslinks are the most prevalent form of cisplatin-induced DNA adducts (>90%) (7) and are bypassed by the one- or the two-Pol mechanism. *In vitro* experiments revealed that Pol β can bypass a d(GpG)-cisplatin intrastrand adduct (141). Additionally, it has been reported that TLS Pol η *in vitro* is able to bypass a d(GpG)-cisplatin intrastrand adduct thereby preferentially incorporates a cytosine over the d(GpG)-cisplatin intrastrand adduct (179). Recently, it has been shown *in vivo*, that either TLS Pol κ or TLS Pol η incorporate the correct nucleotide opposite the d(GpG)-cisplatin intrastrand adduct and TLS Pol ζ carries out the extension step (194). Similarly, *in vivo* experiments proposed a model where RAD18/RAD6 dependent monoubiquitination of PCNA activates the bypass of the d(GpG)-cisplatin intrastrand adduct by TLS Pol η and where activated TLS Pol ζ performs the subsequent extension in a Rev1-dependent manner (122).

4.4.9. Interstrand-crosslink (ICL)

Beside the major intrastrand-crosslinks, cisplatin only forms less than 10% ICLs (7).
It has been found using a plasmid-based recombination assay, that the repair of ICLs in the recombination-dependent repair involves components of mismatch repair, ERCC1-XPF, REV3, Fanconia anemia proteins and homologous recombination (121). Other

findings revealed that ICL are also repaired in a recombination-independent pathway including both NER and TLS Pol ζ and Rev1 (195).

In summary, TLS Pols are involved in tolerating a broad range of DNA damages highlighting the importance of TLS in the maintenance of genomic integrity.
It has been shown that a controlled regulation of the TLS Pols is essential to bypass DNA lesions and to sustain DNA replication. However, a stringent regulation of the TLS Pols is indispensable to avoid the accumulation of mutations.
In combination with the DNA repair mechanisms, the TLS Pols build a powerful mechanism to maintain genomic integrity.

4.5. Relevance of TLS Polymerases in cancer therapy

As reviewed by Lange et al. (65), the TLS Pol are discussed to have an emerging role as targets in cancer therapy, since TLS Pol are involved in both DNA damage repair and tolerance mechanisms.
The TLS Pols contribute to mutagenesis due to the incorporation of incorrect nucleotides opposite DNA lesions. Since the accumulation of mutations can give rise to carcinogenesis and development of resistance, the inhibition of TLS Pols was proposed as a cancer treatment strategy, which should sensitize cancer cells to DNA damaging agents and at the same time reduce damage-induced mutagenesis (196). To date, no specific inhibitors for Y family TLS Pols are available except for the pyrene nucleotide analogs OXT-GTP and OX-ATP, which are able to inhibit TLS Pol η (197). Additionally, some natural inhibitors are known to have inhibitory effects such as Petasiphenol, which is a specific inhibitor of TLS Pol λ *in vitro* [(198, 199)]. It has been shown that Petasiphenol has antiangiogenic activity [(200) and reviewed in (198)]. Tormetic acid is another inhibitor of TLS Pol λ and β but also of the replicative DNA Pols such as Pol α. Tormetic acid showed an antitumorigenic activity *in vivo* [reviewed in (198)].
Another study has shown that gene expression of TLS Pol θ is upregulated in two cohorts of patients with untreated primary breast cancers. Interestingly, the upregulation of gene expression of TLS Pol θ correlated with poor clinical outcome. Therefore, inhibition of the expression/activity of TLS Pol θ may improve the clinical outcome and increase responsiveness to genotoxic treatments (201).
The concept of "synthetic lethality" describes another therapeutical approach where defects in two pathways alone can be tolerated but become lethal when combined. This

concept has been extended by the idea of "synthetic sickness", whereas the combined loss/mutation of function of two genes does not kill cells but significantly impairs cellular fitness (Fig.9).

Gene A	Gene B	
+	+	viable
+	−	viable
−	+	viable
−	−	lethal/Impaired fitness

Fig.9 The Concept of synthetic lethality/sickness explains that a cell can compensate either the loss of gene A or B but the loss of both genes leads either to lethality or impaired fitness. For details see text.

DDR is often abrogated in cancer cells and it was proposed to develop cancer treatments taking advantage of cancer-specific DDR alterations (28).

The principle of synthetic lethality was successfully applied in cancer therapy of patients carrying mutations in BRCA1 or BRCA2, a specific DNA-repair defect. Inhibition of poly(adenosine diphosphate [ADP]-ribose) polymerase (PARP) resulted in synergistic antitumor activity in the treatment of hereditary ovarian- and breast cancer of patients with BRCA mutation (202).

Experiments with yeast deficient in mismatch repair proteins revealed a synthetic lethal effect when either yeast Pol 3 (human ortholog Pol δ) (203) or yeast Pol 2 (human ortholog Pol ε) (204) was inhibited. The overlapping functions of mismatch repair and DNA Pol proofreading activity are conserved from yeast to humans, which indicate that inhibition of specific Pols might selectively kill mammalian tumors deficient in mismatch repair.

Indeed, it was recently shown that inhibition of Pol β or γ induces synthetic sickness/lethality in MSH2-respectively MLH1-deficient human cancer cells (205). Moreover, the inhibition of *Rev3*, the catalytic subunit of TLS Pol ζ, has been shown to sensitize *in vitro* human fibroblasts to cisplatin and decreases the formation of cisplatin resistant cells (112). Recent findings showed that suppression of *Rev3* increases the sensitivity of chemoresistant lung tumors to cisplatin in an *in vivo* mouse xenograft model (117).

As will be documented in chapter 7, inhibition of the TLS Pol ζ *per se* confers synthetic sickness/lethality specifically in a variety of different cancer cells whereas normal cells are less affected, thus identifying REV3 as a potential target for cancer-specific therapy.

5. References

1. Lander ES, Linton LM, Birren B, Nusbaum C, Zody MC, Baldwin J, et al. Initial sequencing and analysis of the human genome. Nature. 2001;409:860-921.
2. Venter JC, Adams MD, Myers EW, Li PW, Mural RJ, Sutton GG, et al. The sequence of the human genome. Science. 2001;291:1304-51.
3. Lindahl T, Barnes DE. Repair of endogenous DNA damage. Cold Spring Harb Symp Quant Biol. 2000;65:127-33.
4. Croteau DL, Bohr VA. Repair of oxidative damage to nuclear and mitochondrial DNA in mammalian cells. J Biol Chem. 1997;272:25409-12.
5. Lindahl T. Instability and decay of the primary structure of DNA. Nature 1993;362:709-15.
6. Lindahl T, Wood RD. Quality control by DNA repair. Science. 1999;286:1897-905.
7. Zamble DB, Lippard SJ. Cisplatin and DNA repair in cancer chemotherapy. Trends Biochem Sci. 1995;20:435-9.
8. Denissenko MF, Pao A, Tang M, Pfeifer GP. Preferential formation of benzo[a]pyrene adducts at lung cancer mutational hotspots in P53. Science. 1996;274:430-2.
9. Harper JW, Elledge SJ. The DNA damage response: ten years after. Molecular cell. 2007;28:739-45.
10. Rouse J, Jackson SP. Interfaces between the detection, signaling, and repair of DNA damage. Science. 2002;297:547-51.
11. Waters LS, Minesinger BK, Wiltrout ME, D'Souza S, Woodruff RV, Walker GC. Eukaryotic translesion polymerases and their roles and regulation in DNA damage tolerance. Microbiol Mol Biol Rev. 2009;73:134-54.
12. Kurz EU, Lees-Miller SP. DNA damage-induced activation of ATM and ATM-dependent signaling pathways. DNA repair. 2004;3:889-900.

13. Bakkenist CJ, Kastan MB. DNA damage activates ATM through intermolecular autophosphorylation and dimer dissociation. Nature. 2003;421:499-506.
14. Kobayashi J, Tauchi H, Chen B, Burma S, Tashiro S, Matsuura S, et al. Histone H2AX participates the DNA damage-induced ATM activation through interaction with NBS1. Biochem Biophys Res Commun. 2009;380:752-7.
15. Stucki M, Jackson SP. MDC1/NFBD1: a key regulator of the DNA damage response in higher eukaryotes. DNA repair. 2004;3:953-7.
16. Zhang J, Powell SN. The role of the BRCA1 tumor suppressor in DNA double-strand break repair. Mol Cancer Res. 2005;3:531-9.
17. Mochan TA, Venere M, DiTullio RA, Jr., Halazonetis TD. 53BP1, an activator of ATM in response to DNA damage. DNA repair. 2004;3:945-52.
18. Lavin MF. The Mre11 complex and ATM: a two-way functional interaction in recognising and signaling DNA double strand breaks. DNA repair. 2004;3:1515-20.
19. Buscemi G, Perego P, Carenini N, Nakanishi M, Chessa L, Chen J, et al. Activation of ATM and Chk2 kinases in relation to the amount of DNA strand breaks. Oncogene. 2004;23:7691-700.
20. Turenne GA, Paul P, Laflair L, Price BD. Activation of p53 transcriptional activity requires ATM's kinase domain and multiple N-terminal serine residues of p53. Oncogene. 2001;20:5100-10.
21. Sengupta S, Harris CC. p53: traffic cop at the crossroads of DNA repair and recombination. Nat Rev Mol Cell Biol. 2005;6:44-55.
22. Deng C, Zhang P, Harper JW, Elledge SJ, Leder P. Mice lacking p21CIP1/WAF1 undergo normal development, but are defective in G1 checkpoint control. Cell. 1995;82:675-84.
23. Di Leonardo A, Linke SP, Clarkin K, Wahl GM. DNA damage triggers a prolonged p53-dependent G1 arrest and long-term induction of Cip1 in normal human fibroblasts. Genes Dev. 1994;8:2540-51.
24. Cortez D, Guntuku S, Qin J, Elledge SJ. ATR and ATRIP: partners in checkpoint signaling. Science. 2001;294:1713-6.
25. Ward IM, Chen J. Histone H2AX is phosphorylated in an ATR-dependent manner in response to replicational stress. J Biol Chem. 2001;276:47759-62.
26. Chini CC, Chen J. Claspin, a regulator of Chk1 in DNA replication stress pathway. DNA repair. 2004;3:1033-7.

27. Gorgoulis VG, Vassiliou LV, Karakaidos P, Zacharatos P, Kotsinas A, Liloglou T, et al. Activation of the DNA damage checkpoint and genomic instability in human precancerous lesions. Nature. 2005;434:907-13.
28. Bartkova J, Horejsi Z, Koed K, Kramer A, Tort F, Zieger K, et al. DNA damage response as a candidate anti-cancer barrier in early human tumorigenesis. Nature. 2005;434:864-70.
29. Halazonetis TD, Gorgoulis VG, Bartek J. An oncogene-induced DNA damage model for cancer development. Science. 2008;319:1352-5.
30. Christmann M, Tomicic MT, Roos WP, Kaina B. Mechanisms of human DNA repair: an update. Toxicology. 2003;193:3-34.
31. Andersen PL, Xu F, Xiao W. Eukaryotic DNA damage tolerance and translesion synthesis through covalent modifications of PCNA. Cell Res. 2008;18:162-73.
32. Chang DJ, Cimprich KA. DNA damage tolerance: when it's OK to make mistakes. Nat Chem Biol. 2009;5:82-90.
33. Schenten D, Kracker S, Esposito G, Franco S, Klein U, Murphy M, et al. Pol zeta ablation in B cells impairs the germinal center reaction, class switch recombination, DNA break repair, and genome stability. J Exp Med. 2009;206:477-90.
34. Rodier F, Coppe JP, Patil CK, Hoeijmakers WA, Munoz DP, Raza SR, et al. Persistent DNA damage signalling triggers senescence-associated inflammatory cytokine secretion. Nat Cell Biol. 2009;11:973-9.
35. Lemontt JF. Mutants of yeast defective in mutation induced by ultraviolet light. Genetics. 1971;68:21-33.
36. Kato T, Shinoura Y. Isolation and characterization of mutants of Escherichia coli deficient in induction of mutations by ultraviolet light. Mol Gen Genet. 1977;156:121-31.
37. Lawrence CW, Das G, Christensen RB. REV7, a new gene concerned with UV mutagenesis in yeast. Mol Gen Genet. 1985;200:80-5.
38. McDonald JP, Rapic-Otrin V, Epstein JA, Broughton BC, Wang X, Lehmann AR, et al. Novel human and mouse homologs of Saccharomyces cerevisiae DNA polymerase eta. Genomics. 1999;60:20-30.
39. McDonald JP, Levine AS, Woodgate R. The Saccharomyces cerevisiae RAD30 gene, a homologue of Escherichia coli dinB and umuC, is DNA damage inducible and functions in a novel error-free postreplication repair mechanism. Genetics. 1997;147:1557-68.

40. Wagner J, Gruz P, Kim SR, Yamada M, Matsui K, Fuchs RP, et al. The dinB gene encodes a novel E. coli DNA polymerase, DNA pol IV, involved in mutagenesis. Molecular cell. 1999;4:281-6.

41. Sharief FS, Vojta PJ, Ropp PA, Copeland WC. Cloning and chromosomal mapping of the human DNA polymerase theta (POLQ), the eighth human DNA polymerase. Genomics. 1999;59:90-6.

42. Aoufouchi S, Flatter E, Dahan A, Faili A, Bertocci B, Storck S, et al. Two novel human and mouse DNA polymerases of the polX family. Nucleic acids research. 2000;28:3684-93.

43. Dominguez O, Ruiz JF, Lain de Lera T, Garcia-Diaz M, Gonzalez MA, Kirchhoff T, et al. DNA polymerase mu (Pol mu), homologous to TdT, could act as a DNA mutator in eukaryotic cells. The EMBO journal. 2000;19:1731-42.

44. Garcia-Diaz M, Dominguez O, Lopez-Fernandez LA, de Lera LT, Saniger ML, Ruiz JF, et al. DNA polymerase lambda (Pol lambda), a novel eukaryotic DNA polymerase with a potential role in meiosis. J Mol Biol. 2000;301:851-67.

45. Marini F, Kim N, Schuffert A, Wood RD. POLN, a nuclear PolA family DNA polymerase homologous to the DNA cross-link sensitivity protein Mus308. J Biol Chem. 2003;278:32014-9.

46. Nelson JR, Lawrence CW, Hinkle DC. Thymine-thymine dimer bypass by yeast DNA polymerase zeta. Science. 1996;272:1646-9.

47. Nelson JR, Lawrence CW, Hinkle DC. Deoxycytidyl transferase activity of yeast REV1 protein. Nature. 1996;382:729-31.

48. Lawrence CW. Cellular roles of DNA polymerase zeta and Rev1 protein. DNA repair. 2002;1:425-35.

49. Lin W, Xin H, Zhang Y, Wu X, Yuan F, Wang Z. The human REV1 gene codes for a DNA template-dependent dCMP transferase. Nucleic acids research. 1999;27:4468-75.

50. Kraemer KH, Slor H. Xeroderma pigmentosum. Clin Dermatol. 1985;3:33-69.

51. Maher VM, Ouellette LM, Curren RD, McCormick JJ. Frequency of ultraviolet light-induced mutations is higher in xeroderma pigmentosum variant cells than in normal human cells. Nature. 1976;261:593-5.

52. Johnson RE, Washington MT, Haracska L, Prakash S, Prakash L. Eukaryotic polymerases iota and zeta act sequentially to bypass DNA lesions. Nature. 2000;406:1015-9.

53. Washington MT, Minko IG, Johnson RE, Wolfle WT, Harris TM, Lloyd RS, et al. Efficient and error-free replication past a minor-groove DNA adduct by the sequential action of human DNA polymerases iota and kappa. Mol Cell Biol. 2004;24:5687-93.

54. Ogi T, Shinkai Y, Tanaka K, Ohmori H. Polkappa protects mammalian cells against the lethal and mutagenic effects of benzo[a]pyrene. Proc Natl Acad Sci U S A. 2002;99:15548-53.

55. Arana ME, Seki M, Wood RD, Rogozin IB, Kunkel TA. Low-fidelity DNA synthesis by human DNA polymerase theta. Nucleic acids research. 2008;36:3847-56.

56. Masuda K, Ouchida R, Takeuchi A, Saito T, Koseki H, Kawamura K, et al. DNA polymerase theta contributes to the generation of C/G mutations during somatic hypermutation of Ig genes. Proc Natl Acad Sci U S A. 2005;102:13986-91.

57. Zan H, Shima N, Xu Z, Al-Qahtani A, Evinger Iii AJ, Zhong Y, et al. The translesion DNA polymerase theta plays a dominant role in immunoglobulin gene somatic hypermutation. The EMBO journal. 2005;24:3757-69.

58. Bertocci B, De Smet A, Weill JC, Reynaud CA. Nonoverlapping functions of DNA polymerases mu, lambda, and terminal deoxynucleotidyltransferase during immunoglobulin V(D)J recombination in vivo. Immunity. 2006;25:31-41.

59. Bertocci B, De Smet A, Berek C, Weill JC, Reynaud CA. Immunoglobulin kappa light chain gene rearrangement is impaired in mice deficient for DNA polymerase mu. Immunity. 2003;19:203-11.

60. Takata K, Shimizu T, Iwai S, Wood RD. Human DNA polymerase N (POLN) is a low fidelity enzyme capable of error-free bypass of 5S-thymine glycol. J Biol Chem. 2006;281:23445-55.

61. Fortune JM, Stith CM, Kissling GE, Burgers PM, Kunkel TA. RPA and PCNA suppress formation of large deletion errors by yeast DNA polymerase delta. Nucleic acids research. 2006;34:4335-41.

62. Kunkel TA. DNA replication fidelity. J Biol Chem. 2004;279:16895-8.

63. McCulloch SD, Kunkel TA. The fidelity of DNA synthesis by eukaryotic replicative and translesion synthesis polymerases. Cell Res. 2008;18:148-61.

64. Boudsocq F, Kokoska RJ, Plosky BS, Vaisman A, Ling H, Kunkel TA, et al. Investigating the role of the little finger domain of Y-family DNA polymerases in low fidelity synthesis and translesion replication. J Biol Chem. 2004;279:32932-40.

65. Lange SS, Takata K, Wood RD. DNA polymerases and cancer. Nat Rev Cancer. 2011;11:96-110.

66. Bienko M, Green CM, Crosetto N, Rudolf F, Zapart G, Coull B, et al. Ubiquitin-binding domains in Y-family polymerases regulate translesion synthesis. Science. 2005;310:1821-4.

67. Guo C, Tang TS, Bienko M, Parker JL, Bielen AB, Sonoda E, et al. Ubiquitin-binding motifs in REV1 protein are required for its role in the tolerance of DNA damage. Mol Cell Biol. 2006;26:8892-900.

68. Avkin S, Sevilya Z, Toube L, Geacintov N, Chaney SG, Oren M, et al. p53 and p21 regulate error-prone DNA repair to yield a lower mutation load. Molecular cell. 2006;22:407-13.

69. Ulrich HD. Deubiquitinating PCNA: a downside to DNA damage tolerance. Nat Cell Biol. 2006;8:303-5.

70. Canitrot Y, Capp JP, Puget N, Bieth A, Lopez B, Hoffmann JS, et al. DNA polymerase beta overexpression stimulates the Rad51-dependent homologous recombination in mammalian cells. Nucleic acids research. 2004;32:5104-12.

71. Bavoux C, Leopoldino AM, Bergoglio V, J OW, Ogi T, Bieth A, et al. Up-regulation of the error-prone DNA polymerase {kappa} promotes pleiotropic genetic alterations and tumorigenesis. Cancer Res. 2005;65:325-30.

72. Friedberg EC, Lehmann AR, Fuchs RP. Trading places: how do DNA polymerases switch during translesion DNA synthesis? Molecular cell. 2005;18:499-505.

73. Lopes M, Foiani M, Sogo JM. Multiple mechanisms control chromosome integrity after replication fork uncoupling and restart at irreparable UV lesions. Molecular cell. 2006;21:15-27.

74. Sarkar S, Davies AA, Ulrich HD, McHugh PJ. DNA interstrand crosslink repair during G1 involves nucleotide excision repair and DNA polymerase zeta. The EMBO journal. 2006;25:1285-94.

75. Waters LS, Walker GC. The critical mutagenic translesion DNA polymerase Rev1 is highly expressed during G(2)/M phase rather than S phase. Proc Natl Acad Sci U S A. 2006;103:8971-6.

76. Niimi A, Brown S, Sabbioneda S, Kannouche PL, Scott A, Yasui A, et al. Regulation of proliferating cell nuclear antigen ubiquitination in mammalian cells. Proc Natl Acad Sci U S A. 2008;105:16125-30.

77. Edmunds CE, Simpson LJ, Sale JE. PCNA ubiquitination and REV1 define temporally distinct mechanisms for controlling translesion synthesis in the avian cell line DT40. Molecular cell. 2008;30:519-29.

78. Brondello JM, Pillaire MJ, Rodriguez C, Gourraud PA, Selves J, Cazaux C, et al. Novel evidences for a tumor suppressor role of Rev3, the catalytic subunit of Pol zeta. Oncogene. 2008;27:6093-101.
79. Karras GI, Jentsch S. The RAD6 DNA damage tolerance pathway operates uncoupled from the replication fork and is functional beyond S phase. Cell. 2010;141:255-67.
80. Seki M, Marini F, Wood RD. POLQ (Pol theta), a DNA polymerase and DNA-dependent ATPase in human cells. Nucleic acids research. 2003;31:6117-26.
81. Seki M, Masutani C, Yang LW, Schuffert A, Iwai S, Bahar I, et al. High-efficiency bypass of DNA damage by human DNA polymerase Q. The EMBO journal. 2004;23:4484-94.
82. Seki M, Wood RD. DNA polymerase theta (POLQ) can extend from mismatches and from bases opposite a (6-4) photoproduct. DNA repair. 2008;7:119-27.
83. Maga G, Shevelev I, Ramadan K, Spadari S, Hubscher U. DNA polymerase theta purified from human cells is a high-fidelity enzyme. J Mol Biol. 2002;319:359-69.
84. Shima N, Munroe RJ, Schimenti JC. The mouse genomic instability mutation chaos1 is an allele of Polq that exhibits genetic interaction with Atm. Mol Cell Biol. 2004;24:10381-9.
85. Shima N, Hartford SA, Duffy T, Wilson LA, Schimenti KJ, Schimenti JC. Phenotype-based identification of mouse chromosome instability mutants. Genetics. 2003;163:1031-40.
86. Ukai A, Maruyama T, Mochizuki S, Ouchida R, Masuda K, Kawamura K, et al. Role of DNA polymerase theta in tolerance of endogenous and exogenous DNA damage in mouse B cells. Genes Cells. 2006;11:111-21.
87. Prasad R, Longley MJ, Sharief FS, Hou EW, Copeland WC, Wilson SH. Human DNA polymerase theta possesses 5'-dRP lyase activity and functions in single-nucleotide base excision repair in vitro. Nucleic acids research. 2009;37:1868-77.
88. Shivapurkar N, Sood S, Wistuba, II, Virmani AK, Maitra A, Milchgrub S, et al. Multiple regions of chromosome 4 demonstrating allelic losses in breast carcinomas. Cancer Res. 1999;59:3576-80.
89. Moldovan GL, Madhavan MV, Mirchandani KD, McCaffrey RM, Vinciguerra P, D'Andrea AD. DNA polymerase POLN participates in cross-link repair and homologous recombination. Mol Cell Biol. 2010;30:1088-96.

90. Yoshimura M, Kohzaki M, Nakamura J, Asagoshi K, Sonoda E, Hou E, et al. Vertebrate POLQ and POLbeta cooperate in base excision repair of oxidative DNA damage. Molecular cell. 2006;24:115-25.

91. Kohzaki M, Nishihara K, Hirota K, Sonoda E, Yoshimura M, Ekino S, et al. DNA polymerases nu and theta are required for efficient immunoglobulin V gene diversification in chicken. The Journal of cell biology. 2010;189:1117-27.

92. Morrison A, Christensen RB, Alley J, Beck AK, Bernstine EG, Lemontt JF, et al. REV3, a Saccharomyces cerevisiae gene whose function is required for induced mutagenesis, is predicted to encode a nonessential DNA polymerase. J Bacteriol. 1989;171:5659-67.

93. Braithwaite DK, Ito J. Compilation, alignment, and phylogenetic relationships of DNA polymerases. Nucleic acids research. 1993;21:787-802.

94. Murakumo Y, Roth T, Ishii H, Rasio D, Numata S, Croce CM, et al. A human REV7 homolog that interacts with the polymerase zeta catalytic subunit hREV3 and the spindle assembly checkpoint protein hMAD2. J Biol Chem. 2000;275:4391-7.

95. Lin W, Wu X, Wang Z. A full-length cDNA of hREV3 is predicted to encode DNA polymerase zeta for damage-induced mutagenesis in humans. Mutation research. 1999;433:89-98.

96. Gibbs PE, McGregor WG, Maher VM, Nisson P, Lawrence CW. A human homolog of the Saccharomyces cerevisiae REV3 gene, which encodes the catalytic subunit of DNA polymerase zeta. Proc Natl Acad Sci U S A. 1998;95:6876-80.

97. Murakumo Y. The property of DNA polymerase zeta: REV7 is a putative protein involved in translesion DNA synthesis and cell cycle control. Mutation research. 2002;510:37-44.

98. Morelli C, Mungall AJ, Negrini M, Barbanti-Brodano G, Croce CM. Alternative splicing, genomic structure, and fine chromosome localization of REV3L. Cytogenet Cell Genet. 1998;83:18-20.

99. Van Sloun PP, Romeijn RJ, Eeken JC. Molecular cloning, expression and chromosomal localisation of the mouse Rev3l gene, encoding the catalytic subunit of polymerase zeta. Mutation research. 1999;433:109-16.

100. Morelli C, Karayianni E, Magnanini C, Mungall AJ, Thorland E, Negrini M, et al. Cloning and characterization of the common fragile site FRA6F harboring a replicative senescence gene and frequently deleted in human tumors. Oncogene. 2002;21:7266-76.

101. Ogawara D, Muroya T, Yamauchi K, Iwamoto TA, Yagi Y, Yamashita Y, et al. Near-full-length REV3L appears to be a scarce maternal factor in Xenopus laevis eggs that changes qualitatively in early embryonic development. DNA repair.9:90-5.

102. Murakumo Y, Ogura Y, Ishii H, Numata S, Ichihara M, Croce CM, et al. Interactions in the error-prone postreplication repair proteins hREV1, hREV3, and hREV7. J Biol Chem. 2001;276:35644-51.

103. Gan GN, Wittschieben JP, Wittschieben BO, Wood RD. DNA polymerase zeta (pol zeta) in higher eukaryotes. Cell Res. 2008;18:174-83.

104. J OW, Kajiwara K, Kawamura K, Kimura M, Miyagishima H, Koseki H, et al. An essential role for REV3 in mammalian cell survival: absence of REV3 induces p53-independent embryonic death. Biochem Biophys Res Commun. 2002;293:1132-7.

105. Esposito G, Godindagger I, Klein U, Yaspo ML, Cumano A, Rajewsky K. Disruption of the Rev3l-encoded catalytic subunit of polymerase zeta in mice results in early embryonic lethality. Curr Biol. 2000;10:1221-4.

106. Wittschieben J, Shivji MK, Lalani E, Jacobs MA, Marini F, Gearhart PJ, et al. Disruption of the developmentally regulated Rev3l gene causes embryonic lethality. Curr Biol. 2000;10:1217-20.

107. Bemark M, Khamlichi AA, Davies SL, Neuberger MS. Disruption of mouse polymerase zeta (Rev3) leads to embryonic lethality and impairs blastocyst development in vitro. Curr Biol. 2000;10:1213-6.

108. Holway AH, Kim SH, La Volpe A, Michael WM. Checkpoint silencing during the DNA damage response in Caenorhabditis elegans embryos. The Journal of cell biology. 2006;172:999-1008.

109. Wittschieben JP, Reshmi SC, Gollin SM, Wood RD. Loss of DNA polymerase zeta causes chromosomal instability in mammalian cells. Cancer Res. 2006;66:134-42.

110. Zander L, Bemark M. Immortalized mouse cell lines that lack a functional Rev3 gene are hypersensitive to UV irradiation and cisplatin treatment. DNA repair. 2004;3:743-52.

111. Krieg AJ, Hammond EM, Giaccia AJ. Functional analysis of p53 binding under differential stresses. Mol Cell Biol. 2006;26:7030-45.

112. Wu F, Lin X, Okuda T, Howell SB. DNA polymerase zeta regulates cisplatin cytotoxicity, mutagenicity, and the rate of development of cisplatin resistance. Cancer Res. 2004;64:8029-35.

113. Li Z, Zhang H, McManus TP, McCormick JJ, Lawrence CW, Maher VM. hREV3 is essential for error-prone translesion synthesis past UV or benzo[a]pyrene diol epoxide-induced DNA lesions in human fibroblasts. Mutation research. 2002;510:71-80.

114. Diaz M, Watson NB, Turkington G, Verkoczy LK, Klinman NR, McGregor WG. Decreased frequency and highly aberrant spectrum of ultraviolet-induced mutations in the hprt gene of mouse fibroblasts expressing antisense RNA to DNA polymerase zeta. Mol Cancer Res. 2003;1:836-47.

115. Wittschieben JP, Patil V, Glushets V, Robinson LJ, Kusewitt DF, Wood RD. Loss of DNA Polymerase {zeta} Enhances Spontaneous Tumorigenesis. Cancer Res.

116. Xie K, Doles J, Hemann MT, Walker GC. Error-prone translesion synthesis mediates acquired chemoresistance. Proc Natl Acad Sci U S A. 2010;107:20792-7.

117. Doles J, Oliver TG, Cameron ER, Hsu G, Jacks T, Walker GC, et al. Suppression of Rev3, the catalytic subunit of Pol{zeta}, sensitizes drug-resistant lung tumors to chemotherapy. Proc Natl Acad Sci U S A. 2010;107:20786-91.

118. Rajpal DK, Wu X, Wang Z. Alteration of ultraviolet-induced mutagenesis in yeast through molecular modulation of the REV3 and REV7 gene expression. Mutation research. 2000;461:133-43.

119. Sonoda E, Okada T, Zhao GY, Tateishi S, Araki K, Yamaizumi M, et al. Multiple roles of Rev3, the catalytic subunit of polzeta in maintaining genome stability in vertebrates. The EMBO journal. 2003;22:3188-97.

120. Ziv O, Geacintov N, Nakajima S, Yasui A, Livneh Z. DNA polymerase zeta cooperates with polymerases kappa and iota in translesion DNA synthesis across pyrimidine photodimers in cells from XPV patients. Proc Natl Acad Sci U S A. 2009;106:11552-7.

121. Zhang N, Liu X, Li L, Legerski R. Double-strand breaks induce homologous recombinational repair of interstrand cross-links via cooperation of MSH2, ERCC1-XPF, REV3, and the Fanconi anemia pathway. DNA repair. 2007;6:1670-8.

122. Hicks JK, Chute CL, Paulsen MT, Ragland RL, Howlett NG, Gueranger Q, et al. Differential roles for DNA polymerases eta, zeta, and REV1 in lesion bypass of intrastrand versus interstrand DNA cross-links. Mol Cell Biol. 2010;30:1217-30.

123. Giannone RJ, McDonald HW, Hurst GB, Shen RF, Wang Y, Liu Y. The protein network surrounding the human telomere repeat binding factors TRF1, TRF2, and POT1. PLoS One. 2010;5:e12407.

124. Matsuoka S, Ballif BA, Smogorzewska A, McDonald ER, 3rd, Hurov KE, Luo J, et al. ATM and ATR substrate analysis reveals extensive protein networks responsive to DNA damage. Science. 2007;316:1160-6.

125. Cantagrel V, Lossi AM, Boulanger S, Depetris D, Mattei MG, Gecz J, et al. Disruption of a new X linked gene highly expressed in brain in a family with two mentally retarded males. J Med Genet. 2004;41:736-42.

126. Kajiwara K, Nagawawa H, Shimizu-Nishikawa S, Ookuri T, Kimura M, Sugaya E. Molecular characterization of seizure-related genes isolated by differential screening. Biochem Biophys Res Commun. 1996;219:795-9.

127. Aravind L, Koonin EV. The HORMA domain: a common structural denominator in mitotic checkpoints, chromosome synapsis and DNA repair. Trends Biochem Sci. 1998;23:284-6.

128. Pfleger CM, Salic A, Lee E, Kirschner MW. Inhibition of Cdh1-APC by the MAD2-related protein MAD2L2: a novel mechanism for regulating Cdh1. Genes Dev. 2001;15:1759-64.

129. Iwai H, Kim M, Yoshikawa Y, Ashida H, Ogawa M, Fujita Y, et al. A bacterial effector targets Mad2L2, an APC inhibitor, to modulate host cell cycling. Cell. 2007;130:611-23.

130. McBride OW, Kozak CA, Wilson SH. Mapping of the gene for DNA polymerase beta to mouse chromosome 8. Cytogenet Cell Genet. 1990;53:108-11.

131. Kumar A, Widen SG, Williams KR, Kedar P, Karpel RL, Wilson SH. Studies of the domain structure of mammalian DNA polymerase beta. Identification of a discrete template binding domain. J Biol Chem. 1990;265:2124-31.

132. Casas-Finet JR, Kumar A, Morris G, Wilson SH, Karpel RL. Spectroscopic studies of the structural domains of mammalian DNA beta-polymerase. J Biol Chem. 1991;266:19618-25.

133. Pelletier H, Sawaya MR, Kumar A, Wilson SH, Kraut J. Structures of ternary complexes of rat DNA polymerase beta, a DNA template-primer, and ddCTP. Science. 1994;264:1891-903.

134. Beard WA, Shock DD, Yang XP, DeLauder SF, Wilson SH. Loss of DNA polymerase beta stacking interactions with templating purines, but not pyrimidines, alters catalytic efficiency and fidelity. J Biol Chem. 2002;277:8235-42.

135. Singhal RK, Wilson SH. Short gap-filling synthesis by DNA polymerase beta is processive. J Biol Chem. 1993;268:15906-11.

136. Prasad R, Beard WA, Wilson SH. Studies of gapped DNA substrate binding by mammalian DNA polymerase beta. Dependence on 5'-phosphate group. J Biol Chem. 1994;269:18096-101.

137. Matsumoto Y, Kim K. Excision of deoxyribose phosphate residues by DNA polymerase beta during DNA repair. Science. 1995;269:699-702.

138. Dianov GL, Prasad R, Wilson SH, Bohr VA. Role of DNA polymerase beta in the excision step of long patch mammalian base excision repair. J Biol Chem. 1999;274:13741-3.

139. Biade S, Sobol RW, Wilson SH, Matsumoto Y. Impairment of proliferating cell nuclear antigen-dependent apurinic/apyrimidinic site repair on linear DNA. J Biol Chem. 1998;273:898-902.

140. Sobol RW, Horton JK, Kuhn R, Gu H, Singhal RK, Prasad R, et al. Requirement of mammalian DNA polymerase-beta in base-excision repair. Nature. 1996;379:183-6.

141. Hoffmann JS, Pillaire MJ, Maga G, Podust V, Hubscher U, Villani G. DNA polymerase beta bypasses in vitro a single d(GpG)-cisplatin adduct placed on codon 13 of the HRAS gene. Proc Natl Acad Sci U S A. 1995;92:5356-60.

142. Esposito G, Texido G, Betz UA, Gu H, Muller W, Klein U, et al. Mice reconstituted with DNA polymerase beta-deficient fetal liver cells are able to mount a T cell-dependent immune response and mutate their Ig genes normally. Proc Natl Acad Sci U S A. 2000;97:1166-71.

143. Gu H, Marth JD, Orban PC, Mossmann H, Rajewsky K. Deletion of a DNA polymerase beta gene segment in T cells using cell type-specific gene targeting. Science. 1994;265:103-6.

144. Horton JK, Srivastava DK, Zmudzka BZ, Wilson SH. Strategic down-regulation of DNA polymerase beta by antisense RNA sensitizes mammalian cells to specific DNA damaging agents. Nucleic acids research. 1995;23:3810-5.

145. Horton JK, Baker A, Berg BJ, Sobol RW, Wilson SH. Involvement of DNA polymerase beta in protection against the cytotoxicity of oxidative DNA damage. DNA repair. 2002;1:317-33.

146. Sobol RW, Watson DE, Nakamura J, Yakes FM, Hou E, Horton JK, et al. Mutations associated with base excision repair deficiency and methylation-induced genotoxic stress. Proc Natl Acad Sci U S A. 2002;99:6860-5.

147. Sobol RW, Prasad R, Evenski A, Baker A, Yang XP, Horton JK, et al. The lyase activity of the DNA repair protein beta-polymerase protects from DNA-damage-induced cytotoxicity. Nature. 2000;405:807-10.

148. Bergoglio V, Pillaire MJ, Lacroix-Triki M, Raynaud-Messina B, Canitrot Y, Bieth A, et al. Deregulated DNA polymerase beta induces chromosome instability and tumorigenesis. Cancer Res. 2002;62:3511-4.

149. Canitrot Y, Laurent G, Astarie-Dequeker C, Bordier C, Cazaux C, Hoffmann JS. Enhanced expression and activity of DNA polymerase beta in chronic myelogenous leukemia. Anticancer Res. 2006;26:523-5.

150. Garcia-Diaz M, Bebenek K, Sabariegos R, Dominguez O, Rodriguez J, Kirchhoff T, et al. DNA polymerase lambda, a novel DNA repair enzyme in human cells. J Biol Chem. 2002;277:13184-91.

151. Yu X, Chini CC, He M, Mer G, Chen J. The BRCT domain is a phospho-protein binding domain. Science. 2003;302:639-42.

152. Ramadan K, Maga G, Shevelev IV, Villani G, Blanco L, Hubscher U. Human DNA polymerase lambda possesses terminal deoxyribonucleotidyl transferase activity and can elongate RNA primers: implications for novel functions. J Mol Biol. 2003;328:63-72.

153. Bebenek K, Garcia-Diaz M, Blanco L, Kunkel TA. The frameshift infidelity of human DNA polymerase lambda. Implications for function. J Biol Chem. 2003;278:34685-90.

154. Picher AJ, Garcia-Diaz M, Bebenek K, Pedersen LC, Kunkel TA, Blanco L. Promiscuous mismatch extension by human DNA polymerase lambda. Nucleic acids research. 2006;34:3259-66.

155. Ma Y, Lu H, Tippin B, Goodman MF, Shimazaki N, Koiwai O, et al. A biochemically defined system for mammalian nonhomologous DNA end joining. Molecular cell. 2004;16:701-13.

156. Terrados G, Capp JP, Canitrot Y, Garcia-Diaz M, Bebenek K, Kirchhoff T, et al. Characterization of a natural mutator variant of human DNA polymerase lambda which promotes chromosomal instability by compromising NHEJ. PLoS One. 2009;4:e7290.

157. Garcia-Diaz M, Bebenek K, Kunkel TA, Blanco L. Identification of an intrinsic 5'-deoxyribose-5-phosphate lyase activity in human DNA polymerase lambda: a possible role in base excision repair. J Biol Chem. 2001;276:34659-63.

158. Zhang Y, Wu X, Yuan F, Xie Z, Wang Z. Highly frequent frameshift DNA synthesis by human DNA polymerase mu. Mol Cell Biol. 2001;21:7995-8006.

159. Gu J, Lu H, Tippin B, Shimazaki N, Goodman MF, Lieber MR. XRCC4:DNA ligase IV can ligate incompatible DNA ends and can ligate across gaps. The EMBO journal. 2007;26:1010-23.

160. Ohmori H, Friedberg EC, Fuchs RP, Goodman MF, Hanaoka F, Hinkle D, et al. The Y-family of DNA polymerases. Molecular cell. 2001;8:7-8.

161. Nair DT, Johnson RE, Prakash S, Prakash L, Aggarwal AK. Replication by human DNA polymerase-iota occurs by Hoogsteen base-pairing. Nature. 2004;430:377-80.

162. Uljon SN, Johnson RE, Edwards TA, Prakash S, Prakash L, Aggarwal AK. Crystal structure of the catalytic core of human DNA polymerase kappa. Structure. 2004;12:1395-404.

163. Silvian LF, Toth EA, Pham P, Goodman MF, Ellenberger T. Crystal structure of a DinB family error-prone DNA polymerase from Sulfolobus solfataricus. Nat Struct Biol. 2001;8:984-9.

164. Trincao J, Johnson RE, Escalante CR, Prakash S, Prakash L, Aggarwal AK. Structure of the catalytic core of S. cerevisiae DNA polymerase eta: implications for translesion DNA synthesis. Molecular cell. 2001;8:417-26.

165. Ohashi E, Murakumo Y, Kanjo N, Akagi J, Masutani C, Hanaoka F, et al. Interaction of hREV1 with three human Y-family DNA polymerases. Genes Cells. 2004;9:523-31.

166. Tissier A, Kannouche P, Reck MP, Lehmann AR, Fuchs RP, Cordonnier A. Co-localization in replication foci and interaction of human Y-family members, DNA polymerase pol eta and REVl protein. DNA repair. 2004;3:1503-14.

167. Guo C, Sonoda E, Tang TS, Parker JL, Bielen AB, Takeda S, et al. REV1 protein interacts with PCNA: significance of the REV1 BRCT domain in vitro and in vivo. Molecular cell. 2006;23:265-71.

168. Sarkies P, Reams C, Simpson LJ, Sale JE. Epigenetic Instability due to Defective Replication of Structured DNA. Molecular cell. 2010;40:703-13.

169. Kannouche P, Stary A. Xeroderma pigmentosum variant and error-prone DNA polymerases. Biochimie. 2003;85:1123-32.

170. Wang Y, Woodgate R, McManus TP, Mead S, McCormick JJ, Maher VM. Evidence that in xeroderma pigmentosum variant cells, which lack DNA polymerase eta, DNA polymerase iota causes the very high frequency and unique spectrum of UV-induced mutations. Cancer Res. 2007;67:3018-26.

171. Jansen JG, Tsaalbi-Shtylik A, Langerak P, Calleja F, Meijers CM, Jacobs H, et al. The BRCT domain of mammalian Rev1 is involved in regulating DNA translesion synthesis. Nucleic acids research. 2005;33:356-65.

172. Delbos F, De Smet A, Faili A, Aoufouchi S, Weill JC, Reynaud CA. Contribution of DNA polymerase eta to immunoglobulin gene hypermutation in the mouse. J Exp Med. 2005;201:1191-6.

173. Ogi T, Kannouche P, Lehmann AR. Localisation of human Y-family DNA polymerase kappa: relationship to PCNA foci. J Cell Sci. 2005;118:129-36.

174. Ogi T, Limsirichaikul S, Overmeer RM, Volker M, Takenaka K, Cloney R, et al. Three DNA polymerases, recruited by different mechanisms, carry out NER repair synthesis in human cells. Molecular cell. 2010;37:714-27.

175. Bebenek K, Tissier A, Frank EG, McDonald JP, Prasad R, Wilson SH, et al. 5'-Deoxyribose phosphate lyase activity of human DNA polymerase iota in vitro. Science. 2001;291:2156-9.

176. Petta TB, Nakajima S, Zlatanou A, Despras E, Couve-Privat S, Ishchenko A, et al. Human DNA polymerase iota protects cells against oxidative stress. The EMBO journal. 2008;27:2883-95.

177. Kannouche P, Fernandez de Henestrosa AR, Coull B, Vidal AE, Gray C, Zicha D, et al. Localization of DNA polymerases eta and iota to the replication machinery is tightly co-ordinated in human cells. The EMBO journal. 2003;22:1223-33.

178. Livneh Z, Ziv O, Shachar S. Multiple two-polymerase mechanisms in mammalian translesion DNA synthesis. Cell Cycle. 2010;9:729-35.

179. Masutani C, Kusumoto R, Iwai S, Hanaoka F. Mechanisms of accurate translesion synthesis by human DNA polymerase eta. The EMBO journal. 2000;19:3100-9.

180. Ramadan K, Shevelev IV, Maga G, Hubscher U. DNA polymerase lambda from calf thymus preferentially replicates damaged DNA. J Biol Chem. 2002;277:18454-8.

181. Maga G, Villani G, Ramadan K, Shevelev I, Tanguy Le Gac N, Blanco L, et al. Human DNA polymerase lambda functionally and physically interacts with proliferating cell nuclear antigen in normal and translesion DNA synthesis. J Biol Chem. 2002;277:48434-40.

182. Zhang Y, Wu X, Guo D, Rechkoblit O, Taylor JS, Geacintov NE, et al. Lesion bypass activities of human DNA polymerase mu. J Biol Chem. 2002;277:44582-7.

183. Choi JY, Lim S, Kim EJ, Jo A, Guengerich FP. Translesion synthesis across abasic lesions by human B-family and Y-family DNA polymerases alpha, delta, eta, iota, kappa, and REV1. J Mol Biol. 2010;404:34-44.

184. Maga G, Villani G, Crespan E, Wimmer U, Ferrari E, Bertocci B, et al. 8-oxo-guanine bypass by human DNA polymerases in the presence of auxiliary proteins. Nature. 2007;447:606-8.

185. Adelman R, Saul RL, Ames BN. Oxidative damage to DNA: relation to species metabolic rate and life span. Proc Natl Acad Sci U S A. 1988;85:2706-8.

186. Belousova EA, Maga G, Fan Y, Kubareva EA, Romanova EA, Lebedeva NA, et al. DNA polymerases beta and lambda bypass thymine glycol in gapped DNA structures. Biochemistry. 2010;49:4695-704.

187. Yoon JH, Bhatia G, Prakash S, Prakash L. Error-free replicative bypass of thymine glycol by the combined action of DNA polymerases kappa and zeta in human cells. Proc Natl Acad Sci U S A. 2010;107:14116-21.

188. Prakash S, Johnson RE, Prakash L. Eukaryotic translesion synthesis DNA polymerases: specificity of structure and function. Annu Rev Biochem. 2005;74:317-53.

189. Kim JK, Patel D, Choi BS. Contrasting structural impacts induced by cis-syn cyclobutane dimer and (6-4) adduct in DNA duplex decamers: implication in mutagenesis and repair activity. Photochem Photobiol. 1995;62:44-50.

190. Yoon JH, Prakash L, Prakash S. Error-free replicative bypass of (6-4) photoproducts by DNA polymerase zeta in mouse and human cells. Genes Dev. 2010;24:123-8.

191. McCulloch SD, Kokoska RJ, Masutani C, Iwai S, Hanaoka F, Kunkel TA. Preferential cis-syn thymine dimer bypass by DNA polymerase eta occurs with biased fidelity. Nature. 2004;428:97-100.

192. Hendel A, Ziv O, Gueranger Q, Geacintov N, Livneh Z. Reduced efficiency and increased mutagenicity of translesion DNA synthesis across a TT cyclobutane pyrimidine dimer, but not a TT 6-4 photoproduct, in human cells lacking DNA polymerase eta. DNA repair. 2008;7:1636-46.

193. Avkin S, Goldsmith M, Velasco-Miguel S, Geacintov N, Friedberg EC, Livneh Z. Quantitative analysis of translesion DNA synthesis across a benzo[a]pyrene-guanine adduct in mammalian cells: the role of DNA polymerase kappa. J Biol Chem. 2004;279:53298-305.

194. Shachar S, Ziv O, Avkin S, Adar S, Wittschieben J, Reissner T, et al. Two-polymerase mechanisms dictate error-free and error-prone translesion DNA synthesis in mammals. The EMBO journal. 2009;28:383-93.

195. Shen X, Jun S, O'Neal LE, Sonoda E, Bemark M, Sale JE, et al. REV3 and REV1 play major roles in recombination-independent repair of DNA interstrand cross-links mediated by monoubiquitinated proliferating cell nuclear antigen (PCNA). J Biol Chem. 2006;281:13869-72.

196. Wang Z. DNA damage-induced mutagenesis : a novel target for cancer prevention. Mol Interv. 2001;1:269-81.

197. Izuta S. Inhibition of DNA polymerase eta by oxetanocin derivatives. Nucleic Acids Symp Ser (Oxf). 2006:269-70.

198. Maga G, Hubscher U. Repair and translesion DNA polymerases as anticancer drug targets. Anticancer Agents Med Chem. 2008;8:431-47.

199. Mizushina Y, Kamisuki S, Kasai N, Ishidoh T, Shimazaki N, Takemura M, et al. Petasiphenol: a DNA polymerase lambda inhibitor. Biochemistry. 2002;41:14463-71.

200. Matsubara K, Mori M, Mizushina Y. Petasiphenol which inhibits DNA polymerase lambda activity is an inhibitor of in vitro angiogenesis. Oncol Rep. 2004;11:447-51.

201. Lemee F, Bergoglio V, Fernandez-Vidal A, Machado-Silva A, Pillaire MJ, Bieth A, et al. DNA polymerase theta up-regulation is associated with poor survival in breast cancer, perturbs DNA replication, and promotes genetic instability. Proc Natl Acad Sci U S A. 2010;107:13390-5.

202. Fong PC, Boss DS, Yap TA, Tutt A, Wu P, Mergui-Roelvink M, et al. Inhibition of poly(ADP-ribose) polymerase in tumors from BRCA mutation carriers. N Engl J Med. 2009;361:123-34.

203. Morrison A, Johnson AL, Johnston LH, Sugino A. Pathway correcting DNA replication errors in Saccharomyces cerevisiae. The EMBO journal. 1993;12:1467-73.

204. Tran HT, Keen JD, Kricker M, Resnick MA, Gordenin DA. Hypermutability of homonucleotide runs in mismatch repair and DNA polymerase proofreading yeast mutants. Mol Cell Biol. 1997;17:2859-65.

205. Martin SA, McCabe N, Mullarkey M, Cummins R, Burgess DJ, Nakabeppu Y, et al. DNA polymerases as potential therapeutic targets for cancers deficient in the DNA mismatch repair proteins MSH2 or MLH1. Cancer cell. 2010;17:235-48.

6. Aim of the thesis

Rev3 is the catalytic subunit of the DNA TLS Pol ζ. It was shown before that inhibition of Rev3 increases the sensitivity of yeast and human normal cells to a variety of DNA damaging agents and reduces the formation of resistant cells. The initial aim of the thesis was to investigate how REV3 inhibition affects sensitivity and resistance formation of cancer cells after cisplatin treatment. Surprisingly, it was found that inhibition of *REV3* expression *per se* suppresses colony formation of a mesothelioma cancer cell line. This finding was subsequently further investigated.

First, the reduction of colony formation after REV3 inhibition was tested for its specificity of cancer cell lines. As a second aim, it was tried to elucidate why colony formation was reduced in cancer cells after REV3 inhibition e.g. induction of apoptosis, senescence or quiescence. The third aim was to address which molecular alteration, e.g. DNA damage accumulation, triggers the reduction in colony formation. The final aim was to identify, which molecular pathway, e.g. DDR, was activated after REV3 inhibition and how the activation of this pathway resulted in the observed reduction of cancer cell growth.

7. Inhibition of *REV3* expression induces persistent DNA damage and growth arrest in cancer cells

Philip A. Knobel, Ilya N. Kotov, Emanuela Felley-Bosco, Rolf A. Stahel and Thomas M. Marti

Laboratory of Molecular Oncology, Clinic and Polyclinic of Oncology, University Hospital Zürich, Häldeliweg 4, CH-8044 Zürich, Switzerland

Neoplasia (2011) 13, 961–970

Corresponding author

Thomas M. Marti

Laboratory of Molecular Oncology, Clinic and Polyclinic of Oncology, University Hospital Zürich, Häldeliweg 4, CH-8044 Zürich, Switzerland

Phone : +41446342874

Fax: +41 44 634 2872

E-mail: thomas.marti@usz.ch

Running Title

REV3 depletion *per se* reduces cancer cell growth

Keywords

REV3, p53, translesion synthesis, persistent DNA damage, cancer

7.1. Abstract

REV3 is the catalytic subunit of DNA translesion synthesis polymerase ζ. Inhibition of *REV3* expression increases the sensitivity of human cells to a variety of DNA damaging agents and reduces the formation of resistant cells. Surprisingly, we found that short hairpin RNA (shRNA)-mediated depletion of *REV3 per se* suppresses colony formation of lung (A549, Calu-3), breast (MCF-7, MDA-MB231), mesothelioma (IL45 and ZL55) and colon (HCT116 +/-p53) tumor cell lines whereas control cell lines (AD293, LP9-hTERT) and the normal mesothelial primary culture (SDM104) are less affected. Inhibition of *REV3* expression in cancer cells leads to an accumulation of persistent DNA damage as indicated by an increase in phospho-ATM-, 53BP1- and phospho-H2AX-foci formation, subsequently leading to the activation of the ATM-dependent DNA damage response cascade. *REV3* depletion in p53-proficient cancer cell lines results in a G_1-arrest and induction of senescence as indicated by the accumulation of p21 and an increase in senescence-associated (SA)-β-Galactosidase activity. In contrast, inhibition of *REV3* expression in *p53*-deficient cells results in growth inhibition and a G_2/M-arrest. A small fraction of the p53-deficient cancer cells can overcome the G_2/M-arrest, which results in mitotic slippage and aneuploidy.

Our findings reveal that *REV3* depletion *per se* suppresses growth of cancer cell lines from different origin whereas control cell lines and a mesothelial primary culture were less affected. Thus, our findings indicate that depletion of REV3 can not only amend cisplatin-based cancer therapy but might also be applied for susceptible cancers as a potential monotherapy.

Abbreviations

TLS, DNA translesion synthesis; Pol ζ, DNA translesion synthesis polymerase ζ; *REV3*, the mammalian *REV3L* gene; MEFs, mouse embryonic fibroblasts; DDR, DNA damage response; DSBs, DNA double strand breaks; ATM, ataxia-telangiectasia mutated; γH2AX, phosphorylated H2AX; P-Chk2, phosphorylated Chk2; AN, aneuploid non-dividing; AD, aneuploid dividing.

7.2. Introduction

Screening in *Saccharomyces cerevisiae* for mutants defective in UV-induced mutagenesis revealed the so called "reversionless" phenotype (REV), which is characterized by a diminished frequency of mutations reverting a specific marker-gene deficiency [1]. Two genes that confer this phenotype when absent are *Rev3* and *Rev7*, the catalytic and the structural subunit of the DNA translesion synthesis (TLS) polymerase ζ (Pol ζ), respectively [2, 3]. The mammalian *REV3L* gene (hereafter *REV3*) encodes a ~350 kDa protein (REV3) consisting of a large C-terminal DNA polymerase subunit, which misses the characteristic proofreading activity present in other B-family DNA polymerases (reviewed in [4]). REV3 interacts via a specific binding domain with REV7 but no additional protein-protein interaction sites were identified. Deletion of *REV3* is embryonic lethal around midgestation [5-8] whereas over-expression of *REV3* leads to increased spontaneous mutation rates [9] confirming that *REV3* expression has to be tightly regulated to maintain genomic integrity. Conversely, one study found that *REV3* expression was down-regulated in colon carcinomas compared to adjacent normal tissue [10] whereas another study found that *REV3* expression was elevated in human gliomas tissues resected before therapy compared to normal brain tissues [11].

Pol ζ belongs to the functional group of TLS DNA polymerases, which are characterized by a less stringent active site and a lower processivity compared to the high fidelity replicative DNA polymerases (reviewed in [4]). TLS polymerases contribute to the maintenance of the genomic integrity by allowing DNA replication to continue in the presence of DNA adducts, which otherwise could lead to DNA replication fork breakdown and subsequent gross chromosomal instability. Pol ζ is the major extender from mismatches formed when incorrect nucleotides are inserted opposite DNA adducts thereby contributing to mutation formation on the nucleotide level. Recently, it was shown that REV3 is involved not only in DNA damage tolerance, but also in DNA repair mechanisms, e.g. interstrand crosslink repair [12-14], homologous recombination [15] and non-homologous end joining as indicated by the deficiency of *REV3*-deleted B cells in class switching of immunoglobulin genes [16].

The unique function of REV3 is highlighted by the fact that the *REV3* depletion increases sensitivity and decreases mutagenesis induced by UV light, cisplatin and other mutagens in human and mouse fibroblasts [15, 17, 18]. In addition, depletion of *REV3* sensitizes mouse B-cell lymphomas and lung adenocarcinomas to cisplatin [19, 20]. Although disruption of mouse *REV3* leads to embryonic lethality, it is possible to generate *REV3*-

deleted mouse embryonic fibroblasts (MEFs) in a p53-deficient background [21]. Spontaneous chromosomal instability was observed in *REV3*-deleted MEFs and *REV3*-deleted cell lines [16, 22, 23].

DNA damage induction results in the activation of an evolutionary conserved signal cascade known as DNA damage response (DDR) (reviewed in [24]). Induction of DNA double strand breaks (DSBs) results in recruitment and activation of ataxia-telangiectasia mutated (ATM). Activated ATM phosphorylates the histone variant H2AX at serine 139 (γH2AX) near DNA DSBs, subsequently leading to an accumulation of DDR proteins at DNA double strand breaks, which can be visualized by immunofluorescence microscopy as distinct foci. Once ATM activation reaches a certain threshold, checkpoint kinase CHK2 is phosphorylated resulting in the accumulation of p53, leading to the accumulation of the cyclin-dependent kinase inhibitor p21. Prolonged activation of p21 after DNA damage is associated with a terminal proliferation arrest, e.g. senescence.

While investigating how inhibition of *REV3* expression affects cisplatin-induced mutagenesis we observed that depletion of *REV3 per se* reduces cancer cell growth whereas growth of control cells is less affected. Suppression of *REV3* expression in cancer cells leads to the accumulation of persistent DNA damage independently of the p53-status. In p53-proficient cancer cells, inhibition of *REV3* expression results in the activation of the ATM-dependent DDR cascade leading to senescence induction. In p53-deficient cancer cells, depletion of *REV3* results in a G_2/M-arrest and increases the fraction of aneuploid cells. In contrast, inhibition of REV3 expression in control cell lines and a mesothelial primary culture neither reduces colony formation nor activates the DDR cascade.

7.3. Material and Methods

Cell Lines and Culture

All cell lines used in this study were authenticated by DNA fingerprinting (Microsynth, Switzerland). SDM104 was maintained as described previously [25]. All other cell lines were maintained in Dulbecco's Modified Eagle's Medium (DMEM) high glucose (Sigma) supplemented with 2 mM L-glutamine, 1 mM sodium pyruvate, 10% fetal calf serum (FCS) and 1% (w/v) penicillin/streptomycin. All cells were grown at 37°C in a humidified atmosphere containing 5% CO_2. Additional details can be found in *Supplemental Materials and Methods*.

Vector Production and Transduction

Replication-deficient lentiviral particles were produced, titrated and used for transduction as described previously [26, 27]. Additional details can be found in *Supplemental Materials and Methods*.

Plasmid Transfection

Cells were transfected using Lipofectamine™ 2000 (Invitrogen) according to the manufacturer's instructions with pSuperior.puro containing either shSCR or three distinct shRNA sequences targeting the *REV3* mRNA. Additional details can be found in *Supplemental Materials and Methods*.

Colony Formation Assay

Crystal violet staining was performed after colonies were visible by eye and the number of colonies was determined by eye applying the same threshold for colony size to all transduced cell lines. The number of colonies obtained by mock treatment was set to 100%.

Quantitative Real-Time PCR

RNA from samples was isolated using RNeasy Mini kit (Qiagen) and reverse transcription was performed on 300 ng RNA (Qiagen QuantiTect® Reverse Transcription protocol). The quantitative expression of *REV3* mRNA was measured by SYBR-Green PCR assay (Applied Biosystems) on a Prism 5700 detection system (SDS, ABI/Perkin/Elmer). Additional details can be found in *Supplemental Materials and Methods*.

Immunofluorescence Microscopy

Immunofluorescence microscopy was essentially performed as described [28]. Details can be found in *Supplemental Materials and Methods*.

Flow Cytometry

Detection of bromodeoxyuridine (BrdU) incorporation in DNA synthesizing cells was carried out using the anti-BrdU antibody (BD Biosciences #555627) according to the manufacturer's instructions. Additional details can be found in *Supplemental Materials and Methods*.

Senescence Associated (SA) Beta-galactosidase Assay

The expression of SA-β-galactosidase was determined by SA-β-galactosidase staining as described [29].

Western analysis

Protein extracts (30μg) were separated by 4-12% SDS–PAGE and transferred onto PVDF membranes. Immunoblotting was performed as described [30]. Details can be found in *Supplemental Materials and Methods*.

ELISA

Cells were washed three times with PBS and serum-free DMEM was added for 24 hours. Conditioned media were filtered and cell number was determined in every experiment by hemocytometer. ELISA was performed using human IL-6 Quantikine ELISA Kit (R&D systems #D6050). The data were normalized to the cell number and reported as fold difference compared to mock treated control.

Statistical analysis

P-values were calculated using the two-tailed Student's t test, were * represents p values <0.05 and ** represents p values <0.01. Error bars represent standard deviations (SD).

7.4. Results

Depletion of *REV3 per se* suppresses colony formation of cancer cells

To study the effect of *REV3* depletion on cisplatin-induced mutagenesis, we established a lentiviral-based system, which allowed us to significantly inhibit *REV3* expression in all cell lines and the primary culture used in this study (Figure S1A and S1B). Inhibition of *REV3* expression did not significantly reduce colony formation of the control cell line AD293 (99% remaining colonies compared to mock treated control), the primary mesothelial culture SDM104 (81%), and the hTERT-immortalized derivative of the mesothelial primary culture LP9 (LP9-hTERT) (98%) (Figure 1A and S2A). Surprisingly, *REV3* depletion *per se* significantly suppressed colony formation of the p53-proficient adenocarcinoma cell line A549 (30%), the p53-deficient adenocarcinoma cell line Calu-3 (57%), the p53-deficient breast cancer cell line MDA-MB231 (47%), the p53-proficient breast cancer cell line MCF-7 (32%), the human mesothelioma cell line ZL55 (27%) and the rat mesothelioma cell line IL45 (4%) compared to mock treated control (Figure 1A and S2A).

In the isogenic p53-proficient and -deficient HCT116 colorectal carcinoma cell lines, there was no significant difference in the reduction of *REV3* expression levels after transduction with an MOI of 170, as used for the cell lines described above, or a MOI of 800 (Figure S1B). However, only the high-titer transduction significantly suppressed colony formation of p53-proficient (49%) and -deficient HCT116 (54%) compared to mock control (Figure 1B and S2B). *REV3* depletion by high-titer transduction did not significantly reduce colony formation of the control cell line AD293 (74%) compared to mock control (Figure 1B and S2B).

Inhibition of *REV3* expression by transduction with three plasmids, one encoding the same siRNA as the lentiviral-based particles plus two plasmids encoding siRNAs targeting alternative sites of the *REV3* mRNA (named REV3-5 and REV3-6) significantly reduced colony formation in the mesothelioma cell line IL45, whereas the control cell line AD293 was not affected (Figure S3). Therefore, we conclude that the observed reduction in colony formation is due to inhibition of *REV3* expression and not due to an unspecific off-target effect of the REV3-4 siRNA. Thus, *REV3* depletion *per se* significantly suppresses colony formation in cancer cell lines whereas colony formation of control cell lines and a primary mesothelial culture is less affected.

Cancer cells accumulate persistent DNA double strand breaks after *REV3* depletion

53BP1- and γH2AX foci formation is regarded as a maker for DNA double strand breaks [28] and a recent study showed that their numbers where increased after persistent DNA damage induction [31]. 7 days after transduction, *REV3* depletion in A549 cells increased the average number of P-ATM and γH2AX foci per cell by a factor of 3.8 and 2.3, respectively, compared to the mock control (Figure 2A). Inhibition of *REV3* expression increased the fraction of A549 cells containing more than two 53BP1 foci to 34% compared to mock (2%) and scrambled (16%) control (Figure 2A). Similarly, *REV3* depletion in MCF7 breast cancer cells increased the average number of γH2AX and 53BP1 foci per cell by a factor of 3.2 and 2.5, respectively, compared to the mock control (Fig. S4). P-ATM foci formation was also elevated in both p53-proficient and -deficient HCT116 cells after *REV3* depletion by a factor of 2.3 and 2.5, respectively, compared to the scrambled control (Figure 2B). In contrast, inhibition of *REV3* expression in the control cell line AD293 did not significantly increase P-ATM, 53BP1- or γH2AX-foci formation compared to the scrambled control (Figure 2A).

DNA double strand breaks, which are not repaired either due to complex DNA modifications or deficiencies in molecular mechanisms result in the formation of persistent DNA double strand breaks (reviewed in [24]). P-ATM foci at persistent DNA double strand breaks are significantly larger in size than the initial foci detectable immediately after damage initiation [32]. Microscopic analysis revealed that the DDR foci induced in *REV3*-depleted cells 7 days after transduction were larger in size compared to the background DDR-foci present in the mock controls (Figure 2A).

Gross chromosomal instability indicated by an elevated number of micronuclei were observed in MEFs with *REV3* deletion [21]. Similarly, the number of micronuclei increased in A549 cells by a factor of 9 after inhibition of *REV3* expression compared to mock control (Figure 2C). Micronuclei formation was not significantly elevated after inhibition of *REV3* expression in AD293 cells (Figure 2C). Thus, inhibition of *REV3* expression induces the formation of persistent DNA double strand breaks and accumulation of gross chromosomal instability in cancer cell lines whereas the control cell line AD293 is significantly less affected ($p < 0.05$ for γH2AX, P-ATM, 53BP1 foci and micronuclei per cell for both A549 and MCF7 versus AD293; numbers of P-ATM foci per cell were not determined in MCF7 cells). In addition, our results indicate that persistent DDR-foci formation after *REV3* depletion is not dependent on the p53 status.

Inhibition of *REV3* expression suppresses proliferation of cancer cells

Since persistent DNA adducts block DNA replication and activate the DDR pathway we investigated whether inhibition of *REV3* expression results in reduced cellular proliferation. Labelling of newly synthesized DNA with BrdU is an established methodology for the assessment of cellular proliferation (reviewed in [33]).

Quantitative analysis of BrdU incorporation revealed that cellular proliferation of A549 cells was reduced by *REV3* depletion to 21% compared to 37% and 38% in mock and scrambled control, respectively (Figure 3, see also figure 2A). Inhibition of *REV3* expression reduced proliferation of p53-proficient HCT116 cells to 25% and in p53-deficient HCT116 cells to a lesser extend to 33%, compared to 41% and 45% in their corresponding scrambled control, respectively (Figure 3). Similarly, *REV3* depletion also reduced proliferation of MCF7 breast cancer cells to 9.2% compared to 17% and 19% in mock and scrambled control, respectively (Figure S4). In contrast, the percentage of replicating cells in the control cell line AD293 and the primary cell culture SDM104 was not diminished by inhibition of *REV3* expression (Figure 3). Thus, *REV3* depletion suppresses cellular proliferation of the analyzed cancer cells whereas proliferation of control cells is not affected.

REV3 depletion activates the DNA damage response pathway in cancer cells

We investigated whether the observed accumulation of persistent DNA double strand breaks in cancer cells results in the activation of the canonical ATM-kinase mediated DNA damage response pathway, which is induced by DNA double strand breaks (reviewed in [24]). As described above, the number of phospho-ATM foci per cell increased after inhibition of *REV3* expression compared to mock and scrambled control in A549, p53-proficient and -deficient HCT116 cells whereas no significant increase occurred in AD293 control cells (Figure 2A and 2B). In A549 cells, *REV3* depletion resulted in increased phosphorylation of the checkpoint kinase Chk2 (P-Chk2) and the accumulation of p53 and the senescence mediator p21 (Figure 4A), which was also observed in MCF7 breast cancer cells but not in the normal mesothelial primary culture SDM104 (Fig. S5). In p53-proficient HCT116 cancer cells, inhibition of *REV3* expression also resulted in an accumulation of p21, which was absent in the p53-deficient isogenic cell line (Figure 4A). Thus, in the analyzed p53-proficient cancer cells, inhibition of *REV3* expression results in the activation of the canonical ATM-dependent DDR pathway.

REV3 depletion induces a G_1-arrest in p53-proficient cancer cells

We tested whether the activation of the DDR pathway and the reduction in BrdU incorporation due to *REV3* depletion change the cell cycle distribution of cancer cells. Depletion of the S-phase after *REV3* depletion, as mentioned above, was accompanied by a significant increase in the fraction of A549 cells in the G_1-phase of the cell cycle to 62% compared to 53% and 51% in the mock and scrambled control, respectively (Figure 3). Similarly, the fraction of p53-proficient HCT116 cells in the G_1-phase increased to 38% after inhibition of *REV3* expression compared to 26% in scrambled control, respectively (Figure 3). In the control cell line AD293 and the primary cell culture SDM104, neither the fraction of cells in S-phase was decreased nor was the fraction of cells in G_1-phase increased after inhibition of *REV3* expression compared to mock and scrambled control (Figure 3). A small but significant increase in the fraction of cells in the G_2-phase was observed in p53-proficient HCT116 cells after *REV3* depletion (23%) compared to mock (17%) and scrambled control (19%). In addition, protein levels of cyclin E, which accumulates during the G_1-phase and is required for the transition from G_1 to S-phase, increased after inhibition of *REV3* expression in p53-proficient- but not in p53-deficient HCT116 cancer cells (Figure 4A). Thus, inhibition of *REV3* expression in the investigated p53-proficient cancer cell lines induces a G_1-arrest, respectively S-phase depletion, whereas the cell cycle distribution of the investigated control cell line and the primary mesothelial culture was not affected.

Inhibition of *REV3* expression induces senescence in p53-proficient cancer cells

Although inhibition of *REV3* expression slightly increased the fraction of Sub-G_1 cells in p53-proficient A549 and HCT116 cells, no significant induction of apoptosis as indicated by an increased fraction of Sub-G_1 cells (Figure 3) or PARP cleavage (Figure 4A) was observed in the remaining control- and cancer cell lines tested in this study.

Since senescence can be induced by persistent DNA damage [31], we investigated whether cells are senescent after *REV3* depletion. Induction of senescence can not be identified by a single marker but is associated with a variety of distinct cellular and molecular changes (reviewed in [34]). Microscopic analysis after crystal violet staining revealed that the morphology of control AD293 cells was not changed 7 days after inhibition of *REV3* expression compared to mock and scrambled control (Figure S6). In the p53-proficient cancer cell lines included in this study, the majority of the colonies were smaller in size after *REV3* depletion and the cells of these colonies displayed morphological changes which are associated with senescence, e.g. increased cell size and flattened shape, whereas cell morphology was not affected in mock and scrambled control (Figure S6). SA-β-Galactosidase staining revealed increased SA-β-Galactosidase activity in IL45, A549- and HCT116 p53-proficient cells after inhibition of *REV3* expression (Figure S6 and Table 1). No increase in SA-β-Galactosidase staining after inhibition of *REV3* expression was detectable in the control cells AD293 or in the p53-deficient MDA-MB231 and HCT116 cancer cell lines. As mentioned above, G_1-arrest, respectively S-phase depletion and p21 accumulation were observed in A549 and p53-proficient HCT116 cells after *REV3* depletion (Figure 3 and 4A).

An increase in persistent DNA damage indicated by residual 53BP1/γH2AX foci is associated in human foreskin fibroblasts with a senescence-associated secretory phenotype including cytokine secretion such as IL-6 [31]. 12 days after transduction, IL-6 secretion was increased in A549 cell after inhibition of *REV3* expression compared to mock and scrambled control (Figure 4B). In contrast, *REV3* depletion in p53-deficient HCT116 cells did not result in a G_1-accumulation nor did it increase p21 levels or increase SA-β-Galactosidase staining (Figure 3, 4A, S6 and Table 1). Similarly, G_1-accumulation and SA-β-Galactosidase staining were abolished in A549 by p53 inhibition (Figure 3 and Table 1). Thus, among the analyzed cancer cell lines, *REV3* depletion *per se* induces senescence in p53-proficient cancer cells only.

REV3 depletion induces a G_2/M arrest and aneuploidy in p53-deficient cancer cells

No G_1-arrest was detectable in p53-deficient HCT116 cell after inhibition of *REV3* expression (Figure 3). Instead, *REV3* depletion in the p53-deficient HCT116 cell line significantly increased the fraction of cells in the G_2/M-phase to 26% compared to 17% and 18% in the mock and scrambled control, respectively (Figure 3). In addition, inhibition of *REV3* expression also increased the fraction of aneuploid cells, which did not incorporate BrdU (AN, aneuploid non-dividing) to 7% compared to 3% and 4% in the mock and scrambled control (Figure 3). The fraction of aneuploid cells, which were still incorporating BrdU (AD, aneuploid dividing) was not increased after *REV3* depletion in p53-deficient HCT116 cells compared to mock and scrambled control (Figure 3). Thus, inhibition of *REV3* expression in the investigated p53-deficient cancer cells results in the accumulation of G_2/M-arrested- and non-dividing aneuploid cells.

In an effort to provide proof-of-principle, we inhibited p53 expression in p53-proficient A549 cancer cells (Figure S1C). Inhibition of p53 expression in A549 cells resulted in a significant increase of the cells in the G_2-phase (22%) and in aneuploidy (total 14%) compared to p53-proficient A549 cells (7.5% and 1.8%, respectively) (Fig. 3), which is in agreement with the dominant role of p53 in the induction of the G_1-arrest [35]. The dominant role of p53 in protection from aneuploidy is highlighted by the finding that additional inhibition of *REV3* expression in combination with p53 inhibition did not further increase aneuploidy in A549 cancer cells.

7.5. Discussion

During our study on the involvement of REV3 in chemotherapy response we found that lentiviral-based inhibition of *REV3* expression was as efficient in the analyzed cancer cell lines as in the primary mesothelial culture and the control cell lines but surprisingly colony formation was reduced in the cancer cell lines only. Therefore, we conclude that reduction in colony formation does not simply mirror the degree of *REV3* expression inhibition relative to scrambled control.

We found that colony formation was not significantly reduced in the control cell lines AD293 and LP9-hTERT and the primary mesothelial culture SDM104 and after inhibition of REV3 expression. This is consistent with previous studies where no deficiency in cell growth/survival was mentioned after antisense-based inhibition of *REV3* expression in human non-tumor cell lines [17, 36]. In contrast, it was shown by different groups that *REV3* knockout reduced cell growth of MEFs [21, 37]. Thus, additional studies will be necessary to clarify how normal cells adapt their DDR to tolerate the loss of REV3 function. At this point it is worth mentioning that investigations of cancer-specific pathways are usually performed using so-called "normal" cells as control. However, normal cells have a limited lifespan [38], which also applies to the primary mesothelial culture SDM104. In contrast, the control cell lines AD293 and LP9-hTERT a virally transformed or immortalized by transfection with human telomerase, respectively, to achieve unlimited proliferation in cell culture. Thus, AD293 and LP9-hTERT might not fully represent normal cells although they have been widely used as normal controls [39, 40] and their response to REV3 depletion was consistent with the reaction of the primary mesothelial cell culture SDM104.

Studies have shown controversial results on the effect of *REV3* depletion on cancer cell growth. On one hand, no deficiency in cell growth/survival was mentioned after si/shRNA-based inhibition of *REV3* expression in HCT116, U2OS and HeLa cancer cells [10, 14, 41]. On the other hand it was shown that knockout of *REV3* resulted in a pronounced growth retardation in Burkitt's Lymphoma cells [42]. We found that inhibition of *REV3* expression *per se* reduced colony formation in lung, breast, mesothelioma and colon tumor cell lines. There are two possible explanations for these apparently controversial observations on the effects of *REV3* depletion in cancer cells. First, the absence of cell growth inhibition in stable cancer cell lines depleted of *REV3* might be due to the genetic modifications acquired during clonal selection, e.g. rewiring of cell-cycle checkpoint pathways [43]. Thus, it would be interesting to identify if the clones isolated in the studies mentioned above acquired genetic modifications compared to their parental cell lines. Secondly, when investigated, it was found that inhibition of *REV3* expression *per se* increased DNA damage levels in cancer cells even when no effect on cell growth/survival was mentioned [10, 14, 42]. Thus, it is possible that the DNA damage level necessary for DDR activation is different in the tested cell lines, explaining the presence or absence of growth arrest (reviewed in [44]).

The second possibility is illustrated by the fact that only inhibition of *REV3* expression by high-titer transduction resulted in a reduction of colony formation in MMR-deficient HCT116 cells although *REV3* expression was not further reduced. It was shown before that activation of the DDR is impaired in MMR-deficient HCT116 cells [45]. Thus, a higher level of cellular stress in form of additional DNA double strand breaks due to more viral integration events after high-titer transduction might be required in HCT116 cells for the induction of a DDR resulting in the reduced colony formation after inhibition of *REV3* expression.

Additionally, the p53 status influences cell-fate after *REV3* depletion. The p53 status did not affect the accumulation of persistent DNA double strand breaks indicated by P-ATM foci after inhibition of *REV3* expression in HCT116 cells. Similarly, a recent study showed that DNA damage accumulation after prolonged activation of the mitotic checkpoint is also independent of the p53 status [46]. Thus, p53 does not protect cancer cells from damage accumulation due to *REV3* depletion although the subsequent cellular outcome, as discussed below, is dependent on the p53-status.

Previously, accumulation of H2AX phosphorylation in U2OS human osteosarcoma cells was observed after REV3 depletion [10]. Microscopic analysis revealed that inhibition of REV3 expression in cancer cells resulted in the accumulation of persistent DNA damage foci, which was also observed after exposure to high-dose ionizing radiation [31], suggesting the accumulation of irreparable DNA double strand breaks. Similarly, the accumulation of large 53BP1 foci was also observed after the induction of mild replication stress or the genetic ablation of the BLM helicase [47]. Interestingly, a very recent publication showed that large 53BP1 foci mark sites of replication stress, which is passed on to daughter cells [48] giving rise to the possibility that the large 53BP1 foci detected after REV3 depletion mark sites of incomplete DNA synthesis rather than DSBs due to replication fork breakdown.

Cellular senescence limits the proliferation of damaged cells that are at risk for neoplastic transformation (reviewed in [34]). Our data indicates that, at least in p53-proficient cancer cells, senescence induction after *REV3* depletion might prevent further transformation of cancer cells by establishing an essentially irreversible growth arrest. It is also proposed that the senescence-associated secretory phenotype, which we observed after inhibition of *REV3* expression indicated by increased IL-6 secretion, might stimulate the immune system to clear senescent cells (reviewed in [34]). However, if senescent cells are not cleared by the immune system, they remain in a "dormant" state representing a dangerous potential for tumor relapse.

A recent study showed that nocodazole (a microtubule polymerization inhibitor) treatment of p53-deficient HCT116 cells leads to prolonged mitosis and subsequent return of the mitotically arrested cells to interphase without cell division resulted in aneuploidy [46], a process known as mitotic slippage. We observed that *REV3* depletion in the p53-deficient HCT116 cell line and in combination with p53 inhibition in the A549 cell line leads to an accumulation of G_2/M-arrested cells and an increase in the frequency of aneuploid cells as it was also described in p53-deficient *REV3*-null MEFs [37].

Based on these results we propose a model (Figure 5) in which inhibition of *REV3* expression can be tolerated in normal cells but results in the accumulation of persistent DNA damage in cancer cells harbouring cancer-specific alterations. Accumulation of persistent DNA damage leads in p53-proficient cancer cells to senescence whereas *REV3* depletion in *p53*-deficient cells results in growth inhibition and a G_2/M-arrest. A small fraction of the p53-deficient cancer cells can overcome the G_2/M-arrest, which results in mitotic slippage and aneuploidy.

The concept of "synthetic lethality", where defects in two pathways alone can be tolerated but become lethal when combined, has been originally described in Drosophila and yeast genetic studies [49, 50]. This concept has been extended by the idea of "synthetic sickness", whereas the combined loss/mutation of function of two genes does not kill cells but significantly impairs cellular fitness [51].

A recent study showed that inhibiting specific DNA repair polymerases induces synthetic sickness/lethality specifically in MMR-deficient cells [52]. In analogy, we found that *REV3* depletion induces synthetic sickness/lethality in the investigated cancer cells. It will be interesting to identify the underlying cancer-specific alteration(s), which render the investigated cancer cell lines prone to growth inhibition due to REV3 depletion. In this context it was shown that DNA repair- and/or cell cycle checkpoint mechanisms are frequently abrogated in cancer cells [53] and the concentration of endogenous DNA damage is higher in human tumoral tissue compared to corresponding adjacent normal tissue (reviewed in [54]). Therefore, differences in repair capacity or DNA damage levels between normal and cancer cells might be the underlying cause for the observed increased sensitivity of cancer cells to *REV3* depletion. Alternatively, replication stress due to the activation of oncogenes might sensitize cancer cells to inhibition of *REV3* expression. A recent study showed that over-expression of Sch9, the *S. cerevisiae* homologue of the mammalian proto-oncogenes Akt and S6, increases superoxide-dependent DNA damage, which subsequently leads to the REV3-dependent formation of point mutations to avoid gross chromosomal rearrangements [55]. However, we can not exclude that the specific genetic or epigenetic alterations underlying the observed synthetic sickness/lethality after *REV3* depletion might differ between the tested cancer cell lines. Indeed, a recent study showed that REV3 deletion in a *S. cerevisiae* strain containing a particular additional chromosome resulted in decreased colony formation [56]. Cancer cells can not only be addicted to oncogenes but also to non-oncogenes (reviewed in [57]). "Non-oncogene addiction" genes are also required for maintenance of the tumorigenic state but are in contrast to oncogenes not functionally altered or mutated. The most prominent example of a "non-oncogene addiction" gene is PARP, which is essential in BRCA-deficient breast cancer cells. Thus, based on the results of our study we propose that *REV3* functions as a "non-oncogene addiction" gene, whose depletion induces synthetic sickness/lethality specifically in the investigated cancer cell lines. Along those lines, we are performing a genome-wide screen to identify essential molecular pathways in cancer cells whose inhibition will further enhance cell killing in combination with REV3 inhibition.

It will be interesting to determine if (1) DNA damage tolerance by REV3-dependent TLS, (2) REV3-dependent DNA repair or (3) a yet-to-be identified function of REV3 is essential for cancer cell growth. Indeed, the size of mammalian REV3 is approximately double the size of the yeast homologue giving rise to the possibility that the non-conserved region of REV3 harbors a yet-to-be identified functional domain, necessary only in higher organisms.

Acknowledgements

We thank Alexandra Graf for her help in the initial experiments. We thank Dr Bert Vogelstein for HCT116 (p53+/+) and HCT116 (p53-/-) cell line. This study was funded by the Cancer League Zurich and the Sassella Foundation to TMM, the Seroussi Foundation and the Foundation for Applied Cancer Research Zürich to RAS.

Conflicts of interest

The authors have declared that no competing interests exist.

7.6. References

[1] Lemontt JF (1971). Mutants of Yeast Defective in Mutation Induced by Ultraviolet Light. *Genetics* **68**, 21-33.

[2] Morrison A, Christensen RB, Alley J, Beck AK, Bernstine EG, Lemontt JF, and Lawrence CW (1989). REV3, a Saccharomyces cerevisiae gene whose function is required for induced mutagenesis, is predicted to encode a nonessential DNA polymerase. *J Bacteriol* **171**, 5659-5667.

[3] Lawrence CW, Das G, and Christensen RB (1985). REV7, a new gene concerned with UV mutagenesis in yeast. *Mol Gen Genet* **200**, 80-85.

[4] Waters LS, Minesinger BK, Wiltrout ME, D'Souza S, Woodruff RV, and Walker GC (2009). Eukaryotic translesion polymerases and their roles and regulation in DNA damage tolerance. *Microbiol Mol Biol Rev* **73**, 134-154.

[5] Bemark M, Khamlichi AA, Davies SL, and Neuberger MS (2000). Disruption of mouse polymerase zeta (Rev3) leads to embryonic lethality and impairs blastocyst development in vitro. *Curr Biol* **10**, 1213-1216.

[6] Esposito G, Godindagger I, Klein U, Yaspo ML, Cumano A, and Rajewsky K (2000). Disruption of the Rev3l-encoded catalytic subunit of polymerase zeta in mice results in early embryonic lethality. *Curr Biol* **10**, 1221-1224.

[7] Wittschieben J, Shivji MK, Lalani E, Jacobs MA, Marini F, Gearhart PJ, Rosewell I, Stamp G, and Wood RD (2000). Disruption of the developmentally regulated Rev3l gene causes embryonic lethality. *Curr Biol* **10**, 1217-1220.

[8] J OW, Kajiwara K, Kawamura K, Kimura M, Miyagishima H, Koseki H, and Tagawa M (2002). An essential role for REV3 in mammalian cell survival: absence of REV3 induces p53-independent embryonic death. *Biochem Biophys Res Commun* **293**, 1132-1137.

[9] Rajpal DK, Wu X, and Wang Z (2000). Alteration of ultraviolet-induced mutagenesis in yeast through molecular modulation of the REV3 and REV7 gene expression. *Mutation research* **461**, 133-143.

[10] Brondello JM, Pillaire MJ, Rodriguez C, Gourraud PA, Selves J, Cazaux C, and Piette J (2008). Novel evidences for a tumor suppressor role of Rev3, the catalytic subunit of Pol zeta. *Oncogene* **27**, 6093-6101.

[11] Wang H, Zhang SY, Wang S, Lu J, Wu W, Weng L, Chen D, Zhang Y, Lu Z, Yang J, *et al.* (2009). REV3L confers chemoresistance to cisplatin in human gliomas: the potential of its RNAi for synergistic therapy. *Neuro Oncol* **11**, 790-802.

[12] Nojima K, Hochegger H, Saberi A, Fukushima T, Kikuchi K, Yoshimura M, Orelli BJ, Bishop DK, Hirano S, Ohzeki M, et al. (2005). Multiple repair pathways mediate tolerance to chemotherapeutic cross-linking agents in vertebrate cells. *Cancer Res* **65**, 11704-11711.

[13] Raschle M, Knipscheer P, Enoiu M, Angelov T, Sun J, Griffith JD, Ellenberger TE, Scharer OD, and Walter JC (2008). Mechanism of replication-coupled DNA interstrand crosslink repair. *Cell* **134**, 969-980.

[14] Hicks JK, Chute CL, Paulsen MT, Ragland RL, Howlett NG, Gueranger Q, Glover TW, and Canman CE Differential roles for DNA polymerases eta, zeta, and REV1 in lesion bypass of intrastrand versus interstrand DNA cross-links. *Mol Cell Biol* **30**, 1217-1230.

[15] Wu F, Lin X, Okuda T, and Howell SB (2004). DNA polymerase zeta regulates cisplatin cytotoxicity, mutagenicity, and the rate of development of cisplatin resistance. *Cancer Res* **64**, 8029-8035.

[16] Schenten D, Kracker S, Esposito G, Franco S, Klein U, Murphy M, Alt FW, and Rajewsky K (2009). Pol zeta ablation in B cells impairs the germinal center reaction, class switch recombination, DNA break repair, and genome stability. *J Exp Med* **206**, 477-490.

[17] Gibbs PE, McGregor WG, Maher VM, Nisson P, and Lawrence CW (1998). A human homolog of the Saccharomyces cerevisiae REV3 gene, which encodes the catalytic subunit of DNA polymerase zeta. *Proc Natl Acad Sci U S A* **95**, 6876-6880.

[18] Diaz M, Watson NB, Turkington G, Verkoczy LK, Klinman NR, and McGregor WG (2003). Decreased frequency and highly aberrant spectrum of ultraviolet-induced mutations in the hprt gene of mouse fibroblasts expressing antisense RNA to DNA polymerase zeta. *Mol Cancer Res* **1**, 836-847.

[19] Xie K, Doles J, Hemann MT, and Walker GC (2010). Error-prone translesion synthesis mediates acquired chemoresistance. *Proc Natl Acad Sci U S A* **107**, 20792-20797.

[20] Doles J, Oliver TG, Cameron ER, Hsu G, Jacks T, Walker GC, and Hemann MT (2010). Suppression of Rev3, the catalytic subunit of Pol{zeta}, sensitizes drug-resistant lung tumors to chemotherapy. *Proc Natl Acad Sci U S A* **107**, 20786-20791.

[21] Wittschieben JP, Reshmi SC, Gollin SM, and Wood RD (2006). Loss of DNA polymerase zeta causes chromosomal instability in mammalian cells. *Cancer Res* **66**, 134-142.

[22] Van Sloun PP, Varlet I, Sonneveld E, Boei JJ, Romeijn RJ, Eeken JC, and De Wind N (2002). Involvement of mouse Rev3 in tolerance of endogenous and exogenous DNA damage. *Mol Cell Biol* **22**, 2159-2169.

[23] Sonoda E, Okada T, Zhao GY, Tateishi S, Araki K, Yamaizumi M, Yagi T, Verkaik NS, van Gent DC, Takata M, *et al.* (2003). Multiple roles of Rev3, the catalytic subunit of polzeta in maintaining genome stability in vertebrates. *EMBO J* **22**, 3188-3197.

[24] d'Adda di Fagagna F (2008). Living on a break: cellular senescence as a DNA-damage response. *Nat Rev Cancer* **8**, 512-522.

[25] Thurneysen C, Opitz I, Kurtz S, Weder W, Stahel RA, and Felley-Bosco E (2009). Functional inactivation of NF2/merlin in human mesothelioma. *Lung Cancer* **64**, 140-147.

[26] Reed SE, Staley EM, Mayginnes JP, Pintel DJ, and Tullis GE (2006). Transfection of mammalian cells using linear polyethylenimine is a simple and effective means of producing recombinant adeno-associated virus vectors. *J Virol Methods* **138**, 85-98.

[27] Salmon P, and Trono D (2007). Production and titration of lentiviral vectors. *Curr Protoc Hum Genet* **Chapter 12**, Unit 12 10.

[28] Marti TM, Hefner E, Feeney L, Natale V, and Cleaver JE (2006). H2AX phosphorylation within the G1 phase after UV irradiation depends on nucleotide excision repair and not DNA double-strand breaks. *Proc Natl Acad Sci U S A* **103**, 9891-9896.

[29] Dimri GP, Lee X, Basile G, Acosta M, Scott G, Roskelley C, Medrano EE, Linskens M, Rubelj I, Pereira-Smith O, *et al.* (1995). A biomarker that identifies senescent human cells in culture and in aging skin in vivo. *Proc Natl Acad Sci U S A* **92**, 9363-9367.

[30] Hopkins-Donaldson S, Ziegler A, Kurtz S, Bigosch C, Kandioler D, Ludwig C, Zangemeister-Wittke U, and Stahel R (2003). Silencing of death receptor and caspase-8 expression in small cell lung carcinoma cell lines and tumors by DNA methylation. *Cell Death Differ* **10**, 356-364.

[31] Rodier F, Coppe JP, Patil CK, Hoeijmakers WA, Munoz DP, Raza SR, Freund A, Campeau E, Davalos AR, and Campisi J (2009). Persistent DNA damage signalling

triggers senescence-associated inflammatory cytokine secretion. *Nat Cell Biol* **11**, 973-979.

[32] Yamauchi M, Oka Y, Yamamoto M, Niimura K, Uchida M, Kodama S, Watanabe M, Sekine I, Yamashita S, and Suzuki K (2008). Growth of persistent foci of DNA damage checkpoint factors is essential for amplification of G1 checkpoint signaling. *DNA repair* **7**, 405-417.

[33] Quinn CM, and Wright NA (1990). The Clinical-Assessment of Proliferation and Growth in Human Tumors - Evaluation of Methods and Applications as Prognostic Variables. *Journal of Pathology* **160**, 93-102.

[34] Collado M, and Serrano M (2010). Senescence in tumours: evidence from mice and humans. *Nat Rev Cancer* **10**, 51-57.

[35] Di Leonardo A, Linke SP, Clarkin K, and Wahl GM (1994). DNA damage triggers a prolonged p53-dependent G1 arrest and long-term induction of Cip1 in normal human fibroblasts. *Genes Dev* **8**, 2540-2551.

[36] Li Z, Zhang H, McManus TP, McCormick JJ, Lawrence CW, and Maher VM (2002). hREV3 is essential for error-prone translesion synthesis past UV or benzo[a]pyrene diol epoxide-induced DNA lesions in human fibroblasts. *Mutation research* **510**, 71-80.

[37] Zander L, and Bemark M (2004). Immortalized mouse cell lines that lack a functional Rev3 gene are hypersensitive to UV irradiation and cisplatin treatment. *DNA repair* **3**, 743-752.

[38] Hayflick L (1965). The Limited in Vitro Lifetime of Human Diploid Cell Strains. *Exp Cell Res* **37**, 614-636.

[39] Tu Y, and Kim JS (2010). Selective gene transfer to hepatocellular carcinoma using homing peptide-grafted cationic liposomes. *J Microbiol Biotechnol* **20**, 821-827.

[40] Hillegass JM, Shukla A, MacPherson MB, Bond JP, Steele C, and Mossman BT (2010). Utilization of gene profiling and proteomics to determine mineral pathogenicity in a human mesothelial cell line (LP9/TERT-1). *J Toxicol Environ Health A* **73**, 423-436.

[41] Lin X, and Howell SB (2006). DNA mismatch repair and p53 function are major determinants of the rate of development of cisplatin resistance. *Mol Cancer Ther* **5**, 1239-1247.

[42] Gueranger Q, Stary A, Aoufouchi S, Faili A, Sarasin A, Reynaud CA, and Weill JC (2008). Role of DNA polymerases eta, iota and zeta in UV resistance and UV-induced mutagenesis in a human cell line. *DNA repair* **7**, 1551-1562.

[43] Reinhardt HC, Aslanian AS, Lees JA, and Yaffe MB (2007). p53-deficient cells rely on ATM- and ATR-mediated checkpoint signaling through the p38MAPK/MK2 pathway for survival after DNA damage. *Cancer cell* **11**, 175-189.

[44] Al-Ejeh F, Kumar R, Wiegmans A, Lakhani SR, Brown MP, and Khanna KK (2010). Harnessing the complexity of DNA-damage response pathways to improve cancer treatment outcomes. *Oncogene* **29**, 6085-6098.

[45] Brown KD, Rathi A, Kamath R, Beardsley DI, Zhan Q, Mannino JL, and Baskaran R (2003). The mismatch repair system is required for S-phase checkpoint activation. *Nat Genet* **33**, 80-84.

[46] Dalton WB, Yu B, and Yang VW (2010). p53 suppresses structural chromosome instability after mitotic arrest in human cells. *Oncogene* **29**, 1929-1940.

[47] Lukas C, Savic V, Bekker-Jensen S, Doil C, Neumann B, Pedersen RS, Grofte M, Chan KL, Hickson ID, Bartek J, *et al.* (2011). 53BP1 nuclear bodies form around DNA lesions generated by mitotic transmission of chromosomes under replication stress. *Nat Cell Biol* **13**, 243-253.

[48] Harrigan JA, Belotserkovskaya R, Coates J, Dimitrova DS, Polo SE, Bradshaw CR, Fraser P, and Jackson SP (2011). Replication stress induces 53BP1-containing OPT domains in G1 cells. *The Journal of cell biology* **193**, 97-108.

[49] Dobzhansky T (1946). Genetics of Natural Populations. Xiii. Recombination and Variability in Populations of Drosophila Pseudoobscura. *Genetics* **31**, 269-290.

[50] Hartman JLt, Garvik B, and Hartwell L (2001). Principles for the buffering of genetic variation. *Science* **291**, 1001-1004.

[51] Kaelin WG, Jr. (2005). The concept of synthetic lethality in the context of anticancer therapy. *Nat Rev Cancer* **5**, 689-698.

[52] Martin SA, McCabe N, Mullarkey M, Cummins R, Burgess DJ, Nakabeppu Y, Oka S, Kay E, Lord CJ, and Ashworth A (2010). DNA polymerases as potential therapeutic targets for cancers deficient in the DNA mismatch repair proteins MSH2 or MLH1. *Cancer cell* **17**, 235-248.

[53] Bartkova J, Horejsi Z, Koed K, Kramer A, Tort F, Zieger K, Guldberg P, Sehested M, Nesland JM, Lukas C, *et al.* (2005). DNA damage response as a candidate anti-cancer barrier in early human tumorigenesis. *Nature* **434**, 864-870.

[54] Croteau DL, and Bohr VA (1997). Repair of oxidative damage to nuclear and mitochondrial DNA in mammalian cells. *J Biol Chem* **272**, 25409-25412.

[55] Madia F, Wei M, Yuan V, Hu J, Gattazzo C, Pham P, Goodman MF, and Longo VD (2009). Oncogene homologue Sch9 promotes age-dependent mutations by a superoxide and Rev1/Polzeta-dependent mechanism. *The Journal of cell biology* **186**, 509-523.

[56] Sheltzer JM, Blank HM, Pfau SJ, Tange Y, George BM, Humpton TJ, Brito IL, Hiraoka Y, Niwa O, and Amon A (2011). Aneuploidy drives genomic instability in yeast. *Science* **333**, 1026-1030.

[57] Luo J, Solimini NL, and Elledge SJ (2009). Principles of cancer therapy: oncogene and non-oncogene addiction. *Cell* **136**, 823-837.

7.7. Figure Legends

Table 1. Induction of senescence after REV3 inhibition is dependent on p53 level

p53	Cell line	mock	shSCR	shREV3	t-test shSCR/shREV3
+	IL45	2.3 +/- 0.7	2.8 +/- 1.0	44.2 +/- 4.0	**
+	A549	5.8 +/- 0.7	6.0 +/- 1.4	16.1 +/- 2.3	**
-	A549 shP53	1.0 +/- 0.3	0.9 +/- 0.3	0.9 +/- 0.3	NS
-	MDA-MB231	0	0	0	N/A
+	AD293 (normal)	0	0	0	N/A

Shown are means (percentage) of senescent cells +/- standard error of the mean. ** $p < 0.01\%$. Abbreviations: NS, not significant; N/A, not applicable. Three independent experiments were analyzed for all cell lines.

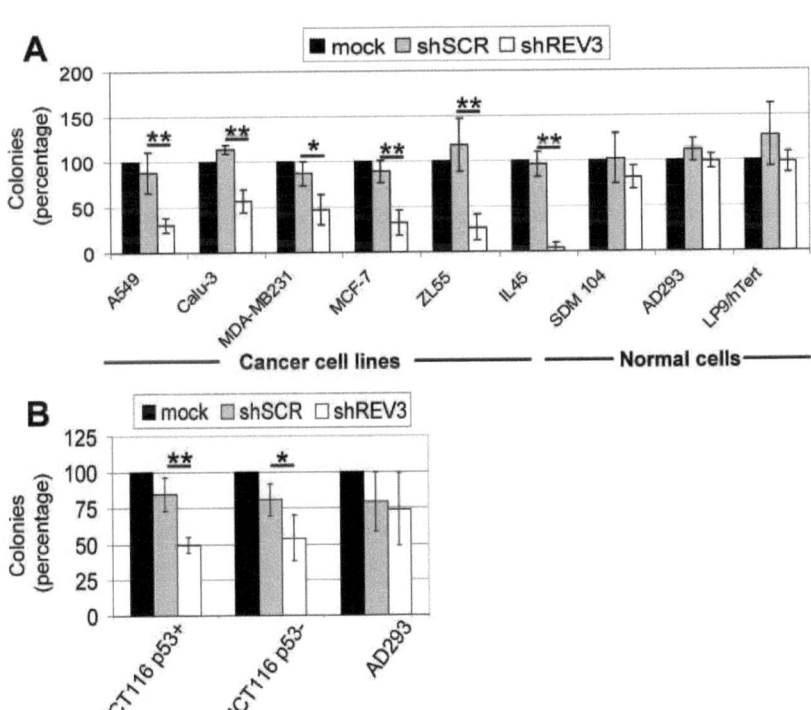

Figure 1. Inhibition of *REV3* expression specifically reduces colony formation of cancer cell lines. Cells were mock treated or transduced with lentiviral-based particles

containing either shSCR or shREV3. (A and B) Cells were stained by crystal violet and total colonies were counted after 2-4 weeks. Colonies were counted from at least 3 independent experiments for all cell lines. Colony number of mock treated cells were set as 100% (*p< 0.05; **p< 0.01). Shown are means +/- standard deviation (SD).

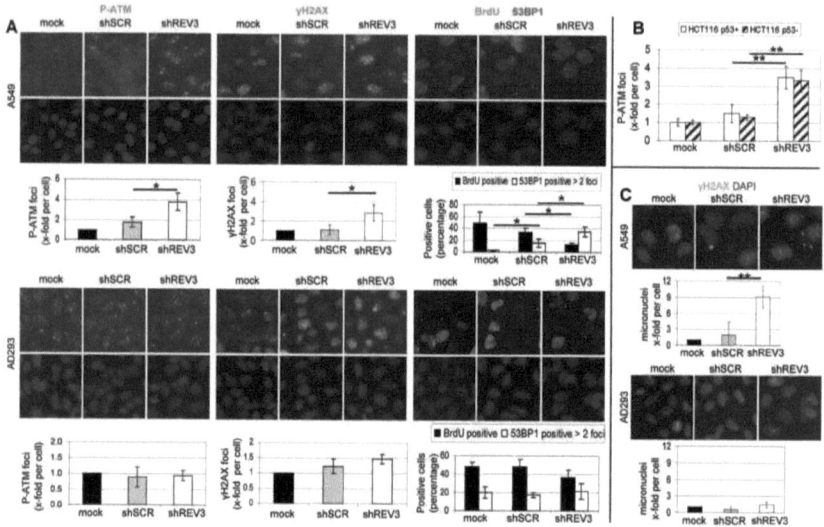

Figure 2. *REV3* depletion induces persistent DNA damage and genomic instability specifically in cancer cells. Cells were mock treated or transduced with lentiviral-based particles containing either shSCR or shREV3 and analyzed after 1 week. (A) Cells were stained for P-ATM, γH2AX or BrdU (all green) and 53BP1 (red) and quantified by immunofluorescence microscopy. Cells containing more than 2 53BP1 foci per cell were considered as 53BP1 positive. (B) Cells were stained for P-ATM and foci per cell were quantified by immunofluorescence microscopy. (C) Cells were stained for γH2AX (green) and nuclear DNA was labelled with DAPI (blue). Micronuclei formation was identified by immunofluorescence microscopy-based analysis of DAPI staining. At least 3 independent experiments were analyzed (*p< 0.05; **p< 0.01). Shown are means +/- SD.

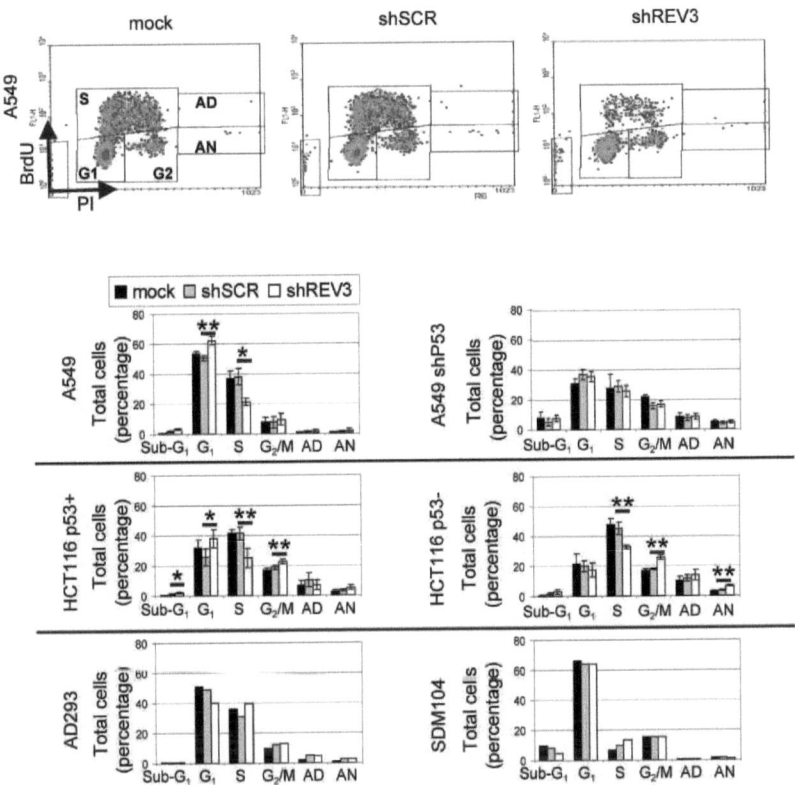

Figure 3. *REV3* depletion changes cell cycle distribution of cancer cell lines. Cells were mock treated or transduced with lentiviral-based particles containing either shSCR or shREV3 and/or lentiviral-based particles containing shP53. After one week, cell cycle distribution was measured by BrdU/PI staining and subsequent FACS analysis. The averages of 3 independent experiments are given for A549, A549 shP53, p53-proficient HCT116 and p53-deficient HCT116 cells whereas representative experiments are shown for SDM104 and AD293 cells (*$p < 0.05$; **$p < 0.01$). (AD: aneuploid dividing cells; AN: aneuploid non dividing cells). Shown are means +/- SD.

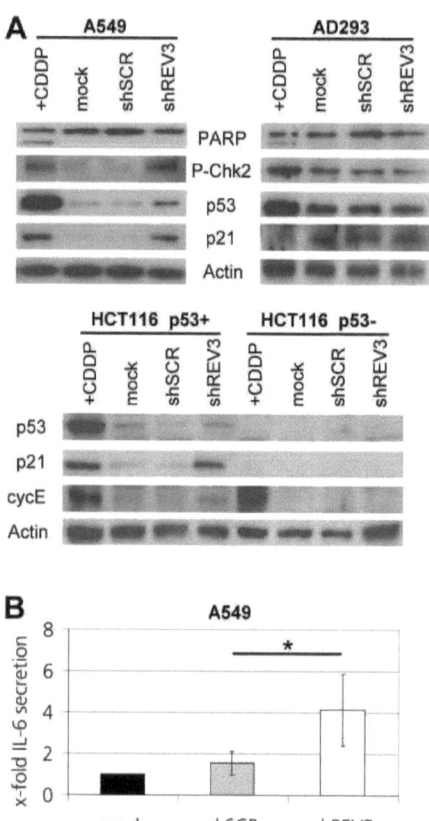

Figure 4. *REV3* depletion induces DNA damage response pathway in cancer cells. Cells were cisplatin- or mock treated or transduced with lentiviral-based particles containing either shSCR or shREV3. (A) After one week, whole cell lysates were analyzed by western analysis. (B) After 24 hours IL-6 secretion in serum-free DMEM was assessed by ELISA, normalized to the cell number and reported as fold increase compared to mock treated control. The averages of at least 3 independent experiments are given. Shown are means +/- SD.

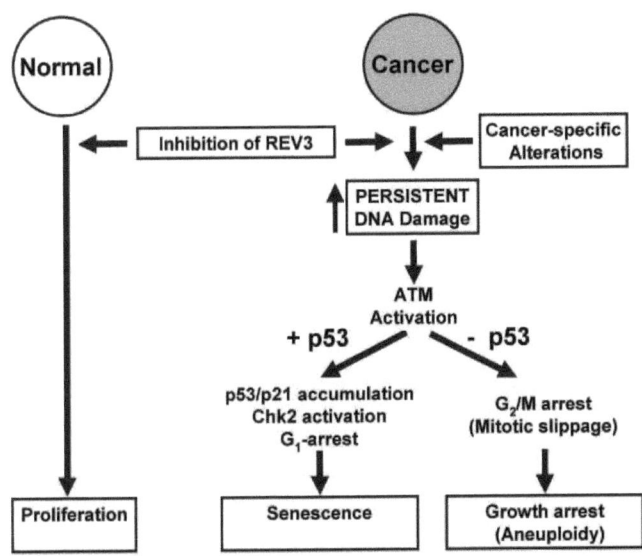

Figure 5. Model: REV3 depletion induces persistent DNA damage specifically in cancer cells, which subsequently results in induction of senescence in p53-proficient- and G_2/M-arrest in p53-deficient cancer cells, respectively **See text for details.**

7.8. Supporting Information

7.8.1. Supplementary Figures

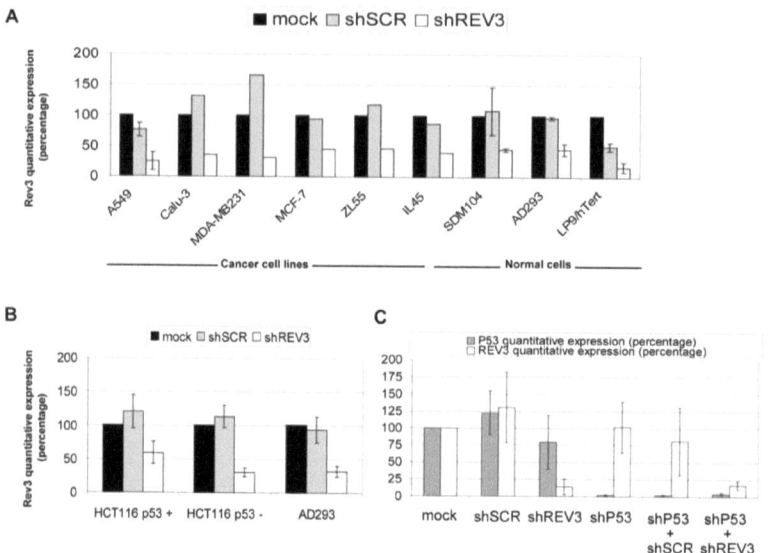

Figure S1. Efficient inhibition of *REV3* expression after transduction with lentiviral-based particles. Cells were mock treated or transduced with lentiviral-based particles containing either shSCR or shREV3 and/or lentiviral-based particles containing shP53. REV3 (A and B) and P53 (C) expression were analyzed by quantitative real time PCR 7 days after transduction. The averages of at least 3 independent experiments are given for A549, IL45, p53-proficient HCT116, p53-deficient HCT116, SDM104 and AD293 cells whereas representative experiments are shown for Calu-3, MDA-MB231, MCF-7, ZL55 and LP9/hTert cells. Rev3 expression levels were normalized to histone H3 expression levels. All Rev3 and p53 expression levels are reported as percentage compared to mock treated A549 (A and C) or p53-proficient HCT116 (B) cells which was set as 100%. Shown are means +/- standard deviation (SD).

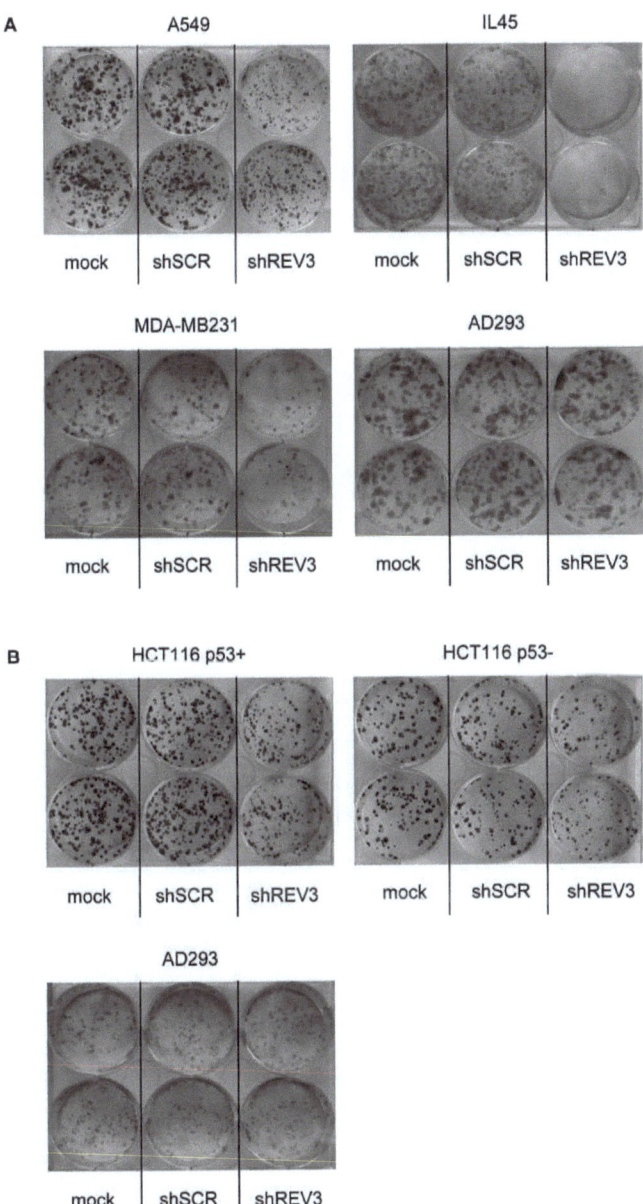

Figure S2. *REV3* **silencing specifically reduces colony formation of cancer cells.**
Cancer cells A549, IL45, MDA-MB231, p53-proficient HCT116, p53-deficient HCT116 and
normal cells AD293 were either mock treated or transduced with lentiviral-based particles

containing shSCR or shREV3. Crystal violet staining was performed once colonies were visible by eye. (A) MOI170 (B) MOI 800. Experiments were made in duplicate wells. See also related Fig.1

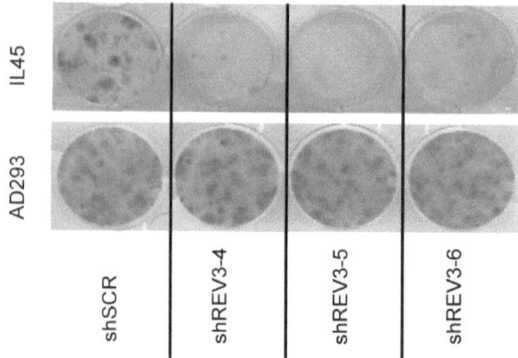

Figure S3. Reduced colony formation after *REV3* silencing is not due to a siRNA off-target effect. IL45 mesothelioma cancer cells and AD293normal cells were either mock treated or transfected with three different plasmids containing shRNA-constructs targeting Rev3. Colonies were stained with crystal violet and counted after 2 weeks. Experiment was made in duplicate wells.

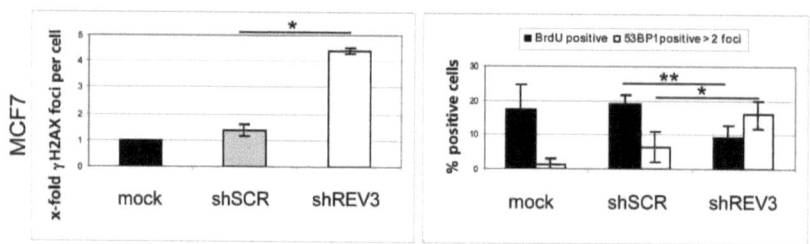

Figure S4. *REV3* depletion induces persistent DNA damage in MCF7 breast cancer cells. Cells were mock treated or transduced with lentiviral-based particles containing either shSCR or shREV3 and analyzed after 1 week. (A) Cells were stained for γH2AX or, BrdU and 53BP1, and quantified by immunofluorescence microscopy. Cells containing more than 2 53BP1 foci per cell were considered as 53BP1 positive. At least 2 independent experiments were analyzed (*$p< 0.05$; **$p< 0.01$). Shown are means +/- SD.

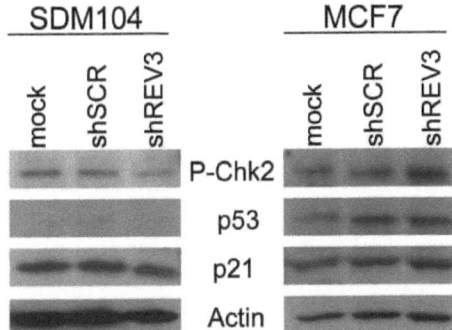

Figure S5. *REV3* depletion induces DNA damage response pathway specifically in cancer cells. Cells were mock treated or transduced with lentiviral-based particles containing either shSCR or shREV3. After one week, whole cell lysates were analyzed by western analysis.

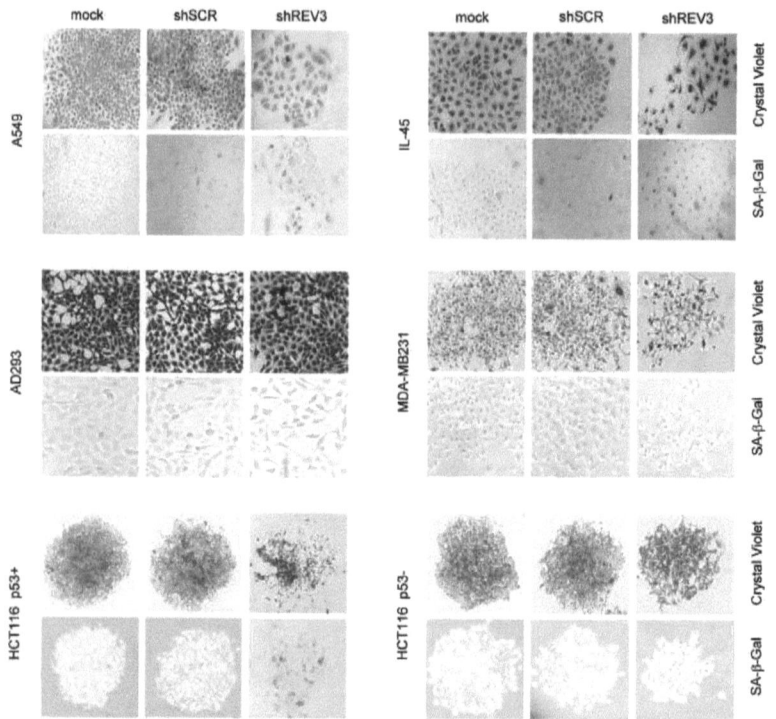

Figure S6. *REV3* silencing induces senescence in p53-proficient cancer cells.
Cancer cells A549, IL45, MDA-MB231, p53-proficient HCT116 and p53-deficient HCT116 and normal cells AD293 were mock treated or transduced with either lentiviral-based particles containing shSCR or shREV3. Crystal violet assay (upper lines) or SA-β-galactosidase assay (bottom lines) was performed after 7 days.

7.8.2 Supplementary Materials and Methods
Cell Lines
The human MPM cell line ZL55 and the primary cell culture SDM104 were generated in our laboratory (25, 30). The rat MPM cell line IL45 was generated elsewhere (craighead et al. 1987). The breast cancer cell lines MDA-MB231 and MCF-7, the adenocarcinoma Calu-3, the squamous NSCLC A549 and the HEK293T were purchased from ATCC. The AD293 cell line, a HEK293 derivative with improved cell adherence, was purchased from Stratagene. The colorectal carcinoma cell lines HCT116 40.16 (p53+/+) and HCT116

379.2 (p53-/-) were kindly provided by Dr. Bert Vogelstein (Johns Hopkins University, Baltimore, MD).

Reagents

When indicated, 20 µM cisplatin (Ebewe) was added for 24 hours.

To clone the short hairpin constructs into the plasmid pSuperior.puro, the following DNA oligonucleotides were ordered from Microsynth:

shREV3-4:

5'-GATCCCCCAAAGATGCTGCTACATTATTCAAGAGATAATGTAGCAGCATCTTTGTTTTA-3'

5'-AGCTTAAAAACAAAGATGCTGCTACATTATCTCTTGAATAATGTAGCAGCATCTTTGGGG-3'

shREV3-5:

5'-GATCCCCGATATTCCATCTGTTACAATTCAAGAGATTGTAACAGATGGAATATCTTTTA-3'

5'-AGCTTAAAAAGATATTCCATCTGTTACAATCTCTTGAATTGTAACAGATGGAATATCGGG-3'

shREV3-6:

5'-GATCCCCTAGTCAGACTTTTCAGCCTTTCAAGAGAAGGCTGAAAAGTCTGACTATTTTTA-3'

5'-AGCTTAAAAATAGTCAGACTTTTCAGCCTTCTCTTGAAAGGCTGAAAAGTCTGACTAGGG-3'

shSCR:

5'-GATCCCCATTCTAGGTGAAAGCTAATTTCAAGAGATAAGATCCACTTTCGATTATTTTTA-3'

5'-AGCTTAAAAATAATCGAAAGTGGATCTTATCTCTTGAAATTAGCTTTCACCTAGAATGGG-3' shP53:

5'-GATCCCCGACTCCAGTGGTAATCTACTTCAAGAGAGTAGATTACCACTGGAGTCTTTTTA-3'

5'-AGCTTAAAAAGACTCCAGTGGTAATCTACTCTCTTGAAGTAGATTACCACTGGAGTCGGG-3'

To quantitatively measure the expression of REV3 mRNA by real time PCR, the following DNA oligonucleotides were ordered from Microsynth:

REV3 forward	5'-TGAGTTCAAATTTGGCTGTACCT-3'
REV3 reverse	5'-TCTAGTCTTCAAAATTTCTTCAAGCA-3'
Histone H3 forward	5'-TAAAGCACCCAGGAAACAACTGGC-3'
Histone H3 reverse	5'-ACCAGGCCTGTAACGATGAGGTTT-3'
P53 forward	5'- GCTTTGAGGTTCGTGTTTGTGCCT-3'
P53 reverse	5'-GCCCACGGATCTTAAGGGTGAAAT-3'

For western analysis, the following primary antibodies were diluted 1:1000: PARP (Cell Signaling #9542), P-Chk2 (R&D Systems #AF1626), p53 (Cell Signaling #9282), p21 (Santa Cruz #sc-756), cyclin E (Santa Cruz #sc247), MAD2B/Rev7 (BD Biosciences #612266). The primary antibody β-Actin (MP Biomedicals #691001) was incubated 1:10'000 for 1 hour at room temperature. The secondary polyclonal antibodies coupled to horseradish peroxidase (HRP) were diluted 1:10'000.

For immunofluorescence microscopy, the following primary antibodies were used: P-ATM 1:1000 (Cell Signaling #sc4526), γH2AX 1:1000 (Upstate #05-636), 53BP1 1:500 (Cell Signalling #4937) and BrdU 1:1000 (BD Biosciences #555627).

Vector Production and Transduction

Short hairpin REV3-4 and scrambled (shSCR) oligos were ligated into pSuperior.puro as described by the manufacturer (OligoEngine). The shRNA and H1 promoter fragments were subsequently ligated into the constitutive expressing lentiviral vector pLVTHM (Addgene). Replication-deficient lentiviral particles were produced and titrated as described previously (28-29).

Cells were seeded in six-well plates (colony formation, immunofluorescence and senescence associated (SA) Beta-galactosidase assay: 500 cells/well [SDM104: 1000 cells/well]; FACS and western analysis: 2500 cells/well [SDM104: 5000 cells/well];

Realtime-polychain reaction (RT-PCR) and ELISA: 5000 cells/well) in 2ml medium. After 6 hours medium was removed and lentivirus suspension added for 30 minutes in 300µl (Immunofluorescence, FACS and RT-PCR multiplicity of infection (MOI) 100; Western analysis and colony formation: MOI 170). All transductions of HCT116- and corresponding control AD293 cells were performed with an MOI800 and incubated for 1 hour. Subsequently, medium was added to final volume of 1.5 ml. For mock treatment, 0.5µm-filtered conditioned medium from a HEK293T culture was added. After 7 days, cells were further processed for distinct experiment as described below except for colony formation, which were incubated longer as described below.

Quantitative Real-Time PCR

Real time PCR cycle conditions were as follows: 1 cycle of 95°C for 10 minutes, 40 cycles of 95°C for 15 seconds and 60°C for 1 minute and 1 cycle of 95°C for 15 seconds and from 60°C slowly elevating to 95°C over several minutes for dissociation curve analysis. *Histone H3* expression was used to standardize the total amount of cDNA and the specificity of the PCR reaction was confirmed by analysis of the melting curve.

Immunofluorescence microscopy

Immunofluorescence microscopy was performed essentially as described before (30). In detail, for BrdU staining, cells were incubated with 10 µM BrdU for 1 hour. Fixation and permeabilization was done with 100% Methanol -20°C and Acetone:Methanol (50:50) -20°C for 20 min at room temperature, respectively. Cells were washed 2x5 min with PBS. For BrdU staining, cells were denatured with 2M HCl for 30min at room temperature and subsequently washed 3x5 min with PBS. PBS containing 1% FBS and 10% BSA was used as blocking solution for 1hour at room temperature. First antibodies were diluted in blocking solution and incubated at 4°C over night. Following antibodies were used: P-ATM 1:1000 (Cell Signaling #sc4526), γH2AX 1:1000 (Upstate #05-636), 53BP1 1:500 (Cell Signalling #4937) and BrdU 1:1000 (BD Biosciences #555627). Cells were washed 2x5 min in PBS. Secondary antibodies were 1:10000 diluted in blocking solution and incubated at 37 °C for 1 hour. Cells were washed 2x5min with PBS and mounted with ProLong Gold Antifade reagent with DAPI (Invitrogen #P36931). Images were acquired with an inverse wide field fluorescence microscope (Leica DM IRBE) equipped with a black and white camera (Hamamatsu ORKA-ER). Image processing with Photoshop (Adobe Systems) was applied to whole images only. Images used for comparison between different transductions were acquired with the same instrument settings and exposure time and were processed equally.

Flow Cytometry

Cell cycle distribution was assed by using a FACSCalibur (FACScan, BD Biosciences, 488 nm excitation laser) and WinMDI software.

Western analysis

Cells were lysed for 30 min on ice in 1× RIPA buffer (Upstate) containing 2xHALT™ protease and phosphatase inhibitor cocktail (Thermo scientific). Cell extracts were denatured at 95°C for 5 minutes, homogenized by successive passing through a 30-gauge syringe needle and protein concentrations were determined using BCA protein assay (Pierce).

9. Acknowledgements

At the end of my PhD there are many people to thank for their support.

First and foremost, I would like to thank Prof. Rolf A. Stahel for the opportunity to work on basic and translational research in his group and his essential support for my PhD thesis.

I would like to express my gratitude to my supervisor Dr. Thomas M. Marti. His enthusiasm supported me throughout my work and I enormously appreciated his advices, his patience in answering all my questions and his help with the experiments. His friendly character always led to a friendly and motivating atmosphere in the lab.

I am particularly grateful to Prof. Ulrich Hübscher for his support, advices and discussions during my PhD and of course being part of my thesis committee.

I would like to thank Prof. Michael Hengartner for kindly accepting to be a member of my thesis committee, for the discussions and his valuable suggestions during the progress reports.

A special thank goes to Dr. Emanuela Felley-Bosco for her discussions, advices and friendly support.

My sincere thanks go to Ilya Kotov for contributing to my work and for his friendship.

A special thank to all the lab members for their helpful discussions and the nice time in and around the lab.

Last but not least I thank my wife Marcela and my family for their limitless support and reliance.

Die VDM Verlagsservicegesellschaft sucht für wissenschaftliche Verlage abgeschlossene und herausragende

Dissertationen, Habilitationen, Diplomarbeiten, Master Theses, Magisterarbeiten usw.

für die kostenlose Publikation als Fachbuch.

Sie verfügen über eine Arbeit, die hohen inhaltlichen und formalen Ansprüchen genügt, und haben Interesse an einer honorarvergüteten Publikation?

Dann senden Sie bitte erste Informationen über sich und Ihre Arbeit per Email an *info@vdm-vsg.de*.

Sie erhalten kurzfristig unser Feedback!

VDM Verlagsservicegesellschaft mbH
Dudweiler Landstr. 99   Telefon +49 681 3720 174
D - 66123 Saarbrücken   Fax   +49 681 3720 1749
www.vdm-vsg.de

Die VDM Verlagsservicegesellschaft mbH vertritt

Printed by Books on Demand GmbH, Norderstedt / Germany